your **allotment**

your allotment

Clare Foster

Photography by Francesca Yorke

CASSELL ILLUSTRATED

contents

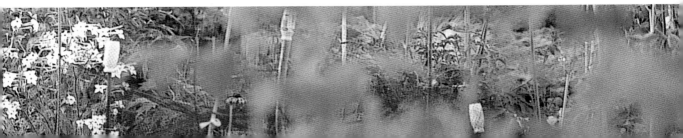

foreword

Do you have an allotment? Are you planning to get one? Do you have space in your garden to grow fruit and vegetables? If so, you need this book.

I've been in love with veg for over 20 years. My 'relationship' with them started as a trader in Covent Garden market where I learned, through the London chefs I supplied, how to cook this huge and exciting range of food. From there the love affair continued through nine years of co-hosting the radio programme *Veg Talk*. Thanks to fruit and vegetables I've experienced many things and met many varied and fascinating people, and it is for this reason, as well as for the many health benefits you will enjoy, that I urge you to join me in growing and enjoying fresh fruit and vegetables.

There's nowhere you can do all this better than on an allotment, where people really are brought together by veg. One of the loveliest things I ever got to do was to present a series about allotments where we visited plots and their growers in Newcastle, Liverpool, Birmingham, London and Bristol. As well as allowing me to sample some of the best produce I'd ever eaten, it convinced me that working your own allotment is a virtual guarantee of a long and healthy life. Consider the combination of light exercise, fresh air, and fresh vegetables and then you'll realise why some of the best allotment growers continue to enjoy tending their plots right through to their eighties.

With easy reference and common sense instructions this book will leave you in no doubt about how to plan and manage your allotment. It covers absolutely everything, from the history of allotments and how to get hold of one, to allotment etiquette and fun ways to encourage children to grow their own.

This book is like having a very friendly and extremely knowledgeable grower standing beside you as you work.

Gregg Wallace

introduction

Allotments are fashionable again. There are approximately 300,000 working allotment plots in Britain today, many of them on thriving sites with long waiting lists, and while this figure can never be compared to the heyday of the allotment during the two world wars, there can be no doubt that their popularity is on the increase.

Why is this, when people lead such busy lives? The main driving force, of course, is the desire to grow healthy, tasty organic food for the family. Barely a week goes by nowadays without another food scare being reported in the press, and people increasingly want to take control over what they eat. Growing your own fruit and vegetables means that you have the reassurance of knowing exactly where your food is coming from: it is super-fresh, packed with nutrients and vitamins and, above all, as flavoursome as some of the supermarket varieties are tasteless.

Allotments are the answer for many people who make the decision to grow their own. No longer the reserve of the archetypal cloth-capped old man, they are attracting a much wider following. Nowadays you'll find people of all ages, backgrounds and cultures on an allotment site, all of them growing different crops in different ways, eager to learn and experiment and, more often than not, determined to grow organically.

Taking on an allotment plot is a commitment and a challenge, but it is also rewarding, educational and thoroughly enjoyable. If you have already made the commitment to take on an allotment, I hope this book will help you through the seasons ahead, giving you tips and advice that will enable you to keep on top of your plot. If you are wavering, then I hope I can persuade you to take the leap. Despite many people's preconceptions, renting an allotment and growing a few basic vegetables really isn't difficult to do, and with a bit of pre-planning and preparation, you could be harvesting your own crops within a few months of starting out.

Clare Foster

all about allotments

From their **beginnings in the 19th century** and through their **heyday during the First and Second World Wars,** allotments have always been part of the **fabric of British society.** Today, although fewer in number, they are much in demand, particularly as the **organic food movement** gathers momentum.

a brief history of the allotment

'Allotments have proved a great boon to many hundreds of thousands of men – particularly the unemployed – in providing them with an interesting spare-time occupation and enabling them to enjoy home-grown supplies of vegetables … During the period of national depression there is not a doubt that allotments were the salvation of many. Even now they are doing a great national service and are to be seen in every town and village throughout the country.' *The Smallholder Encyclopaedia, 1937*

The creation of allotments as we know them today was a direct result of the Enclosure Acts of the 18th and 19th centuries, by which the countryside, more significantly common land, was portioned up and enclosed within fences or hedges. This enclosure of land had been going on since medieval times, as farming methods evolved from the open-field system, and as a result much of the peasant population suffered immeasurably, no longer able to graze sheep or grow crops on the commons. There were rebellions and demonstrations, including the 1649 Diggers invasion of common land near Walton-on-Thames in Surrey, during which angry protestors dug the soil and planted crops to make their point. This was followed by at least nine other Digger colonies setting up camp in various parts of the country over the following year. Tensions were running high, and understandably so.

The years went by, and more of the countryside became out of bounds. Unrest was inevitable, and finally local councils began 'allotting' pieces of land to the poor. It wasn't until 1845 that this became law, with the General Enclosure Act stipulating that land should be set aside as 'field gardens' for the 'labouring population'. These field gardens soon became known as allotments. Although these first allotments were largely in the country, the emphasis quickly shifted away from the provinces towards the cities, simply because of the dramatic expansion of the urban population. By 1906 it became law for local authorities to provide allotment space for the poor, whether in the city or countryside.

The true ascendancy of the allotment came with the two world wars. After the outbreak of World War I in 1914, demand for allotments soared as people worried about feeding their families in wartime Britain. The government sent out an appeal to private garden owners to keep all surplus vegetable seedlings, which the Royal Horticultural Society helped distribute to allotment-holders, while in 1916 local authorities were given the power to create allotments from common land, parks and playing fields. Even the king made it known that potatoes would replace geraniums in the flower beds outside Buckingham Palace and in the royal parks.

After the outbreak of World War II in 1939 people immediately swung into action with the Dig for Victory campaign, and by the end of the war the number of plots had increased to a phenomenal 1,500,000. Countless books and manuals were published as guides to managing a vegetable plot, and, of course, many of the methods expounded in them are just as relevant today.

Inevitably, after the war demand for allotments dropped considerably, and the increased prosperity and lower food prices of the 1950s and 1960s led more and more people to forget about their plots. Since then, there have been peaks and troughs in the popularity of allotments. The 1970s saw a brief return to the land with an increased awareness of ecology, perhaps also helped by the TV comedy series *The Good Life*. And today, once again, we are seeing a surge in demand for plots. In the 1990s gardening was hailed as Britain's favourite pastime, and this, coupled with an increased interest in organic food, has meant that a wider cross-section of people than ever before are taking on allotments. Unfortunately, though, the trend in recent years has been the dwindling number of allotment sites, as many of the city plots are on prime land that is much in demand from residential or commercial developers. Tales of rumoured pay-offs to plot-holders or scaremongering about the threat of development in order to make people vacate their plots are all too frequently heard, and, of course, once the land has been swallowed up there is much less chance of finding new spaces for sites.

But as long as the demand for allotments remains, local authorities are bound by law to find provision. One clause in the Smallholding and Allotment Act of 1908 is still relevant today: if there is sufficient demand for allotments (and in 1908, this was deemed to be a demand in writing from six or more rate-payers), the local authority is bound by law to provide allotments for them. So if enough of us are persistent in our quests to find allotment plots, perhaps we will see new sites opening up, with more and more people taking on an allotment as a way of life.

No matter what happens in years to come, the 21st-century allotment has moved on. It has shaken off its dowdy, elderly image and instead has emerged as a lively, multicultural place where people can go to escape from the pressures of modern life and to produce their own delicious, healthy food.

'There are many reasons for renting an allotment – not least of which is to grow fresh produce. There are the health benefits of a place you can escape to unwind in peace and quiet, with some gentle exercise away from everyday pressures. Conversely, allotment sites are sociable places, great for meeting like-minded people. There is a growing trend towards a wider mix of cultures and ages alongside the traditional plot holders. Tips on growing and cultivating abound making this an ideal environment for the novice gardener.'
Royal Horticultural Society, 2006

how to find an allotment

Most allotment sites are hidden away behind houses and buildings or next to railway lines and rivers, away from the general hubbub of the town or city. Don't let this invisibility put you off – wherever you are, there will be allotments nearby. Some 85 per cent of allotment sites are owned by local authorities; the remaining 15 per cent are privately owned. So the first step in hunting down an allotment is a phone call to the amenity or leisure department of your local council to find out where the nearest allotment sites are. Alternatively, look on your local authority website, do a general internet search for allotments in your area or visit your local library. Then you can either visit the allotment site – although most are under lock and key and not accessible to the general public – or ring the secretary of the local allotment association to find out how to apply for a plot. If you're lucky enough to have a choice of local sites, check that the site you are interested in has a water supply (most do, but some don't), how close you can park your car and find out about additional facilities, such as toilets or a communal meeting space.

Once you have been accepted on to a site, a member of the committee will show you the vacant plot or plots. You may have a choice of several, in which case it would be wise to have a close look at the plot to see what condition it is in. Check for the most common weeds, such as bindweed, couch grass, ground elder and horsetail (see page 158), and ask a few questions about the history of the plot. Some people want to know if the plot was managed organically in previous years, while others are more interested in knowing who their closest neighbours are going to be. Other things to bear in mind are position and aspect. Which way does the plot face? How far is it from the nearest water supply? How far away do you have to park your car?

allotment sizes

🍃 Most allotment plots are 10 poles in size, an archaic measurement that is about 250 square metres (2,700 sq ft), usually in a configuration of 10 x 25 m (33 x 82 ft).

🍃 This is a good size and can be quite a daunting amount of space for a novice gardener, so many associations offer half-size or even quarter-size plots.

🍃 Alternatively, sharing an allotment with a friend or relative is a good idea if you want to share the load, and this can be extremely useful if one or the other is away during the summer when lots of watering is needed.

'When I see sights like the sun going down over all the allotments, catching the Pennines in the distance, it makes me feel alive and acutely aware that I'm on a planet, working with and trying to tame nature.'
Richard Markin, Kiveton Park

allotment regulations

Since they were established 150 years ago, most allotment sites have been governed by their own quirky sets of rules and regulations, enforced with varying degrees of stringency depending on the allotment committee and personnel. On most sites frequent inspection rounds take place, so the most important rule is that you must keep your plot as tidy and weed-free as possible. By law, the committee must give a year's warning before evicting a plot-holder, and this is usually done with a series of warnings to get the plot in shape. Other rules vary from site to site. Some associations are very strict about what you can and can't grow: at the Fulham Palace Allotments in London, for example, plot-holders are allowed to devote only 10 per cent of their plot to flowers, with the emphasis on edible crops. Other rules may prohibit the construction of sheds or of boundaries around the plot, or planting trees or large fruit bushes. If you feel that sticking to stringent rules and regulations might be a problem, it is wise to ask for a copy of the rule book (if one exists) before applying for an allotment or to find out from existing plot-holders how the site is run.

cost
Rent paid for allotment plots varies throughout the country, ranging from £10 to £75 a year. Rents are usually collected at the beginning of each year, and plot-holders can be evicted if rents are in arrears.

security
Allotments can be targets for theft and vandalism, and it is each plot-holder's shared responsibility to deter such attacks as far as possible. If your site has a boundary fence or hedge, see that it is well maintained and report any weaknesses to the appropriate person. If you are allowed a shed, make sure it is securely locked when you aren't there, and never leave expensive equipment on the allotment.

allotment etiquette

The main thing to remember is that your allotment plot is on communal land, and behaviour must therefore be tempered accordingly.

don't
encroach on to **neighbouring land** with your plants, children or animals;

let your plot become overrun with weeds that invade **neighbouring plots**;

plant hedges or trees that might shade a **neighbouring plot**;

have large bonfires at the busiest times (there may be **rules restricting bonfires**);

leave rubbish or piles of weeds **lying around**;

monopolize the **communal water supply**;

steal other people's **prize vegetables**;

blast out loud music when others are trying to enjoy the **peace and quiet**.

do
participate in allotment life;

share excess plants, seeds or vegetables;

be considerate if using chemicals;

keep your plot looking **neat and tidy**.

using the allotment

Obviously the main reason for taking on an allotment is to grow your own fruit and vegetables, but in addition to this, there are a host of other plus-points. Digging, weeding and turning compost are all good exercise, burning calories and keeping you fit, and just being out in the fresh air can increase your sense of well-being. There is a great sense of satisfaction in growing and tending plants – from seeing the first seeds sprout to harvesting the crops you have nurtured – and this connection with nature can be immensely beneficial and rewarding. Allotments also provide invaluable lessons in sustainability, encouraging recycling and resourcefulness. They are the 'green lungs' of the inner city landscape, harbouring wildlife and maintaining biodiversity in a world that is disappearing under concrete.

Allotments can help your social life, too. People chat, make friends, swap seeds and plants, and above all exchange gardening information and ideas. There is a real sense of community on almost all the allotment sites I have visited, and this gives an added dimension to the pleasure of gardening, making it a sociable rather than a solitary pastime. Then there is the creativity of allotment gardening. While many outsiders can't see beyond the higgledy-piggledy patchwork of the allotment site, with its cobbled-together sheds, compost heaps and rows of rectangular plots, there is so much interest here if you look deeper. An allotment can be a place of free expression, whether this is demonstrated in neat, military rows of leeks or in abundant tangles of flowers and pumpkins.

Allotments are, above all, places of recreation. They are used in different ways by different people, from the retired couple who spend most days on the allotment growing an impressive array of vegetables, to the busy family who manage to get there at weekends only. Some people use the allotment as the hub of their social life, cooking the fruits of their labour on portable gas stoves and enjoying the open-air way of life. For those who live in blocks of flats, they are substitute gardens, somewhere they can come in their free time to be outside and sit in a deckchair – to rest, of course, from the hours of gardening just done!

'In the high-pressure, deadline-driven world we live in your allotment gives you an opportunity to relax, and exercise your mind and hands. The peace and tranquillity also help to remove tension, reduce stress and give you time to think. You are your own boss, work at the same pace as nature, and like other forms of moderate exercise gardening will help to lower blood pressure and cholesterol levels.' *Carl Stratham, Halifax*

children and allotments

More and more young families are discovering the joys of having an allotment. For children of all ages, especially those who are being brought up in an inner city, an allotment plot can be an exciting playground, full of new experiences, sounds, sights and smells. All children love digging around in the dirt, and older children enjoy helping to plant seed and harvest crops. Some parents may feel a certain level of frustration at having to slow things down to accommodate little hands dabbling around in the seedbed, but this is more than made up for by their excitement at seeing things grow, and knowing that they are learning about good food, about nature, and playing healthily outdoors.

ideas for a child-friendly plot

Give your children **their own patch of land** to cultivate by making small, square raised beds for them, and give them **easy, quick crops** to grow, such as sunflowers or dwarf beans, so that their interest is sustained.

Grow **interesting, brightly coloured vegetables**, such as multicoloured chard or purple beans, to attract their attention.

Give them **large, brightly coloured labels** on which they can write their own name as well as the name of the plant.

Buy **child-sized tools** so that they can copy you when you are digging or watering.

Get a worm bin; as well as providing a quick and easy method of composting, worms are a constant source of fascination for children.

Start a family pumpkin or marrow **competition;** if you have room, grow a pumpkin for each child, carving their initials into the skin when the fruit is small.

Devote an area of the plot to **wildlife-friendly** plants to attract butterflies and bees.

Give your children old bits of material, such as wood or netting, for them to **construct dens** or other structures.

All children love digging holes or making mounds, so **let them get as dirty as they like!**

Keep visits to the allotment short or break the day up with a picnic.

the first steps

You've handed over your first rent cheque and are now officially an allotmenteer. Depending on the time of year and state of your allotment, you may be taking home your first crops within a few months. But don't rush. It pays to put in the ground-work now – clearing and digging, eradicating weeds, improving the soil and laying out the plot. Time spent preparing the ground will make life much easier when you start the growing process.

tools of the trade

It pays to invest a little in your tools. You will be giving them good, hard use on the allotment and buying the best quality you can afford is worthwhile. You can buy on-line, but going to a shop and handling the tools for their feel and weight before buying is advisable. Nowadays there are endless permutations on the basic tool palette: for an allotment, you needn't buy more than a few essentials, acquiring others as you go on as and when you need them.

tools

spade It is useful to have both full-size and border-size, with a steel head and wooden or plastic shaft. Choose from either a D- or T-shaped handle.

fork Again, forks are available in large and small (border) sizes. The tines should be made from a single piece of steel for strength.

trowel Needed for planting and potting up, transplanting seedlings and numerous other tasks.

hand-fork Useful for hand-weeding and many other tasks.

garden reel and line Necessary for making those super-straight drills.

secateurs and shears Needed for pruning and clearing away plant material.

hoe There are two kinds: draw hoe and Dutch hoe, both used for weeding. You can also use a draw hoe to mark out drills. If you are going for one only, a Dutch hoe is the more useful.

rake You will use a rake for levelling and working the soil to a fine tilth in preparation for sowing seeds. Look for one with a forged steel head.

watering can Choose from either metal or plastic.

wheelbarrow Essential for moving compost, manure, weeds, grass clippings or spent vegetable matter.

gardening gloves Have plenty of these around – tough, thick ones for tackling prickly weeds, and thinner ones for lighter tasks.

'There are many forms of false economy in gardening, and the purchase of cheap tools cannot be too strongly condemned. They are inefficient in use, trying to the temper and, of course, expensive in the long run. Buy the best tools you can afford and choose them with care'.

The Vegetable Garden Displayed, 1942

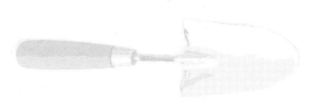

clearing the plot

Surveying your very own plot of land for the first time can be daunting, particularly if it is hidden under a tangle of couch grass and nettles. The unfortunate reality is that, nine times out of ten, when a plot is handed over it is overgrown and neglected. The degree of neglect, however, is just down to luck.

If you are fortunate enough to inherit a plot that has been empty for only a short time, the weeds may just be superficial, mainly annuals, and easy to deal with. If your plot has been left untended for several years, you will have a much more arduous task in front of you, especially if some of the more pernicious perennial weeds, such as brambles, bindweed or ground elder, have made it their home.

clearing an overgrown plot

The first step in clearing an overgrown plot is to scythe, chop or strim the weeds down to ground level, which can be hard work. Leave the debris in the sun for several weeks to die off, or under some old carpet or tarpaulin, where it will rot down, then add it to the compost heap at a later stage when you are sure that the weeds are well and truly dead.

The second step: Once you have chopped back the top of the weeds, the roots must be tackled – and this is where the back-breaking work begins. Many perennial weeds have unbelievably large root systems, some of them penetrating several metres into the soil, and most have the ability to grow from small sections of rhizome (underground stem

sections). There is no easy way to eradicate these toughies organically, other than digging them out with a fork – and continuing to dig them out again and again as they inevitably reappear. If you are tenacious enough, eventually the root system will weaken and the weeds will disappear altogether, although this can take many years. I always think that controlling is a better word to use when dealing with perennial weeds, because it is possible to cultivate a plot that has perennial weeds, as long as they don't get out of hand.

In the initial stages of clearing a plot many people resort to using rotavators, without realizing that they are doing more long-term harm than good. All that a rotavator does is chop into fine pieces the roots and rhizomes of the worst offenders, and because many perennial weeds multiply so persistently from the tiniest section of root, you are spreading the problem in the process.

weeds

Weeds are simply unwanted plants, usually natives that are best adapted to the soil and conditions in the local area, and because they compete with cultivated plants for nutrients, water and light, they must constantly be controlled. Like other plants, weeds can be divided into three categories: annuals, biennials and perennials.

Annual weeds germinate and mature in a season and die away, leaving a copious supply of seed that will germinate the following year. They don't penetrate lower than the topsoil and are easily removed by hoeing or pulling them out by hand. If annual weeds are left to grow for too long, they set seed, creating problems the following year.

Biennials take two seasons to produce seed and then die back. In the first season they germinate in early summer and form a low rosette of leaves with a long taproot. The following season they produce a stem and flower, which sets seed. The best method of removal is uprooting, preferably in the first year before it gets too big.

Perennials are the most persistent and problematic weeds, and they survive from season to season by a highly efficient underground root system. To eradicate perennial weeds, the entire root system must be destroyed, and this can be done either by digging them out or suppressing them under a layer of plastic or other material. The main offenders when it comes to perennial weeds are bindweed, horsetail, couch grass, ground elder and brambles – the likelihood is that you will encounter at least one of these when you take on an allotment (see page 158 to identify them).

covering the soil

An alternative to digging weeds out by hand, inevitably a long, painstaking process, is to cover the soil with a layer of thick black plastic, old carpet or thick cardboard, which will starve the weeds of light and eventually kill them. The disadvantage of this method, apart from the aesthetics, is that it can take several seasons for the weeds to die – even up to three years with some weeds. You can, however, plant vegetables through the plastic sheeting by making slits in the fabric. If you decide to do this, it's best to buy special porous membrane so that water and air aren't excluded, thereby keeping the soil organisms alive and the soil in good condition. As well as suppressing the weeds, covering the soil like this (technically called mulching) has the added advantage of retaining moisture in the soil, which is especially valuable on an allotment when you may not be able to water as often as you might like.

Covering the soil with material such as old carpet can help to suppress weeds, and crops can be successfully planted and grown through it.

'I don't think it's a case of to cover up or not cover up. Local conditions are the most important factor – i.e., climate and soil type. We don't all get snow in the winter and frosts can be few and far between for others. Some have heavy clay soil that is close to being waterlogged in winter, whereas others have light sandy soil that can be dug over at any time. Part of the joy of learning about horticulture is learning from those who have decades of experience and learning from their mistakes as well as their successes.'

Martin Parkes, Hemel Hempstead

An alternative is to cover your plot in winter only, which will protect the soil from hard frosts and drying winds as well as suppressing the weeds. Other materials, such as leaves or manure, can be used for winter cover instead of plastic, or you could try planting a green manure, which as well as acting as a weed-suppressant will also add goodness to the soil when it is dug back in (see page 61). If you don't want to cover all your plot, you could start off by cultivating half, and covering half with a weed suppressant until a later stage – a good compromise if it's too daunting a task to clear your plot in one go.

laying out the plot

Once you have cleared your plot you have an exciting blank canvas to work with, and the next step is to come up with a design that is going to work for you. Look at neighbouring plots to see how they are laid out and talk to other plot-holders about how to make the best use of the space. A trial-and-error approach is fine, but in the long run it pays to divide the plot into logical, organized spaces that are easily accessible and easy to tend.

Make a list of what you want to grow and think about how to arrange the space, factoring in room for compost bins, sheds or greenhouses. Your plot can be as simple as a series of symmetrical rectangular beds and mown grass paths; or as complex as a parterre, with diagonal dividing lines and gravel paths. Beds can be raised and contained within wooden planks or edged with brick or stone, while paths can be grass, brick or bark chip.

Traditionally, allotment plots were designed around long rows of vegetables running north to south across the rectangular plot (see opposite). Some people still find this the easiest way to manage their plot, but others prefer to be individual, designing their space to suit their needs and even to reflect their personality. Plots can be decorative, the beds edged with flowers and herbs like a French potager; they can even be non-productive, acting more like a garden where families come at weekends to relax and enjoy the outdoors.

'It is good practice to have, when possible, the vegetable rows running from north to south rather than from east to west, for thus only can the sun shine down between the rows, ripening the crops evenly'.
The Smallholder Encyclopaedia, 1937

MINISTRY OF AGRICULTURE AND FISHERIES' WAR-TIME ALLOTMENT PLAN

Compost Heap	Tool Shed	Seedbed
	Tomatoes, Marrow, Radish, Parsley	

C

MISCELLANEOUS CROPS

DWARF PEAS .. .. 3 rows 2 ft. 6 in. apart.	Intercrop with Spinach.. 2 rows and follow with Leeks .. .. .. 4 rows 1 ft. apart.
DWARF BEANS .. 2 rows 2 ft. 6 in. apart.	
ONIONS .. .. .. 8 rows 1 ft. apart.	Follow with Spring Cabbage .. 4 rows
SHALLOTS .. .. 2 rows 1 ft. apart BROAD BEANS .. 1 double row	Follow with Winter Lettuce Intercrop with Summer Lettuce
RUNNER BEANS .. 1 row	

A

POTATOES AND ROOT CROPS

PARSNIP .. .. .. 3 rows 15 in. apart.	
CARROT (MAINCROP) 5 rows 1 ft. apart.	
POTATOES (EARLY) .. 3 rows 2 ft. by 1 ft.	Follow with Turnips
POTATOES (OTHERS) 6 rows 2 ft. by 1 ft. 3 in.	
SPINACH BEET or SEAKALE BEET .. .. .. 1 row	

B

WINTER AND SPRING GREEN CROPS

	Maincrops CABBAGE (WINTER) 3 rows 2 ft. by 2 ft.
Intercrop space for Savoys and Brussels Sprouts with Early Carrots 2 rows, and Early Beet, 1 row.	SAVOYS .. .. .. 2 rows 2 ft. by 2 ft. BRUSSELS SPROUTS 2 rows 2 ft. 6 in. by 2 ft. 6 in.
Early Dwarf Peas .. 1 row	SPROUTING BROCCOLI 2 rows 2 ft. by 2 ft. KALE .. .. .. 2 rows 2 ft. by 2 ft. SWEDES .. .. 2 rows 1 ft. 3 in. apart. GLOBE BEET .. 2 rows 1 ft. 3 in. apart.

Rotation Diagram

C	B	A
A	C	B
B	A	C

NOTE.—This Cropping Plan is not drawn to scale. The ground dimensions of the whole plot are 30 ft. by 90 ft. Sections A, B and C are each 30 ft. by 28 ft., and the space provided for Seedbed, Tool Shed, etc., is 30 ft. by 6 ft.

paths

Paths should be carefully thought-out at the planning stage, and should follow a logical route through the plot, giving the easiest access to beds, seating area, compost bins and the entrance/exit to your plot. Don't skimp on width: the main axis path should be wide enough to accommodate a wheelbarrow, while subsidiary paths can be slightly narrower. The easiest option for paths on the allotment is simply to leave them as bare earth. The soil will soon be trodden down and compacted, and weeds will find it hard to thrive if you are regularly using each pathway. Many allotment paths are turfed, but they will need regular mowing and edging to make sure the grass doesn't encroach into the beds. Bark chips are an alternative, but this can be an expensive and labour-intensive option, especially if your allotment is a long way from your car. Old pavers, reclaimed bricks, gravel or even decking could also be used. If your plot is full of stones, put them to good use when you dig them out by using them on your pathways.

Clockwise from top left: Options for paths include grass, wooden planking or decking, paving stones or bark chips.

sheds and structures

Many allotmenteers are lucky enough to inherit sheds or greenhouses from previous occupants, but if you have an empty plot you'll probably want to erect some sort of structure to house all your tools (provided allotment rules allow this). Imaginative plot-holders with lots of time often make fantastic sheds with reclaimed materials, but if you're short of time there are plenty of proprietary kits available.

The shed will be the hub of your plot, probably with a seating area in front of it, so site it near the entrance to your plot – at the top of the plot rather than the bottom. For serious growers, a greenhouse may also be necessary, and most allotment sites allow them. Some plot-holders opt for mini greenhouses made from plastic, while others, in true allotment spirit, rig up their own from plastic sheeting. The only other structure you will almost certainly need is a compost bin (or several), and it is sensible to draw in an area for compost on your initial plan.

Top row: Sheds, greenhouses and other structures can be home-made, like this mini polytunnel made from plastic sheeting (left) or built from a kit like this shed (opposite).

Bottom row: Recycled materials and imagination were used to build this shed (left). Plastic greenhouses like this one (opposite) are widely available to buy, and can house tender plants and seedlings.

boundaries and windbreaks

Many allotments have rules on boundaries, including what you may or may not erect or plant, or the height of the finished structure or planting – you could be casting shade over a neighbouring plot if your boundary is too high. Traditionally, allotments are not enclosed spaces, but often there will be a need to create shelter from the prevailing wind, and this can be done by:

- planting fruit bushes, such as raspberries, gooseberries or currants, which can provide some shelter for other crops.

- planting a row of Jerusalem artichokes, which are perennial and look good too.

- planting a hedge (check your allotment rules first).

- creating a row of espaliered or cordon apple or pear trees (see page 152).

- erecting a low, natural-looking fence – woven willow, for example.

'I have just started planting a wild rose hedge (*Rosa rugosa*). It will have a magnificent scent in the summer and will attract a lot of beneficial insects.'

E. Jones, Nottingham

Top row: Espaliered apples make an attractive boundary or screen on the plot (left). A row of artichokes makes a striking and effective windbreak (opposite).

Bottom row: Reclaimed materials form the boundary of this plot (left). A rose hedge is grown along one end of an allotment to screen it from the neighbouring plot (opposite).

drawing up a plan

When you are designing the plot, one of the most important things to bear in mind is access, both generally – where do you enter and exit the plot? – and specifically – how are you going to get round the plot to tend your crops? Aspect is also very important. Which way does the plot face? Are there any trees, walls or other structures that might throw shade on the plot? Where does the prevailing wind come from, and is there any existing shelter? These are all questions that you should ask yourself when you are thinking about the design.

making a scale drawing

Once you have decided how you would like to lay out the plot, sketch out the design on paper, before marking it out on the plot. If your design is complex, it is sensible to make a scale drawing so that you can make exact measurements, resulting in a cohesive, neat-looking space. You'll need a long, steel measure (available from builders' merchants), as well as stakes and twine to mark out the beds.

Measure the boundaries of your plot and then transfer them to a piece of graph paper. Usually a scale of 1:50 is the ideal, with each square on the graph paper representing a foot or a metre. Now mark on any existing features, such as trees or plants you don't want to get rid of, and measure any existing structures to transfer them accurately to your plan. Once you have the outline plan, you can start to develop the overall layout, thinking carefully about paths, access and the size of the beds. When the plan is drawn up, you can transfer it to the plot, marking out the paths and beds with stakes and twine or a can of garden spray paint (available from most builders' merchants).

1. basic grid with raised beds

This simple layout is easy to construct and maintain and can be adapted to any size plot. It is designed around a symmetrical grid pattern, with 12 self-contained raised beds and a central path. Each bed has pathways all the way round for easy access. At the top of the plot is the shed, with a seating area just in front of it, and a space for herbs, flowers or tomatoes, which can be grown in pots. In the opposite corner, also for easy access, are the compost bins, a cold frame and a seedbed. Right down at the far end of the plot is a long, narrow bed, which can be used for a number of different purposes. With a simple layout like this, crop rotation is easy: you can simply divide the plot into three or four sections and make the crops rotate downwards towards the end of the plot (see page 49).

Making raised beds enclosed within wood or brick takes a bit of extra time and effort, but there are many advantages to them, not least because they create a good cohesive structure that lends itself to crop rotation. They also improve drainage, help to deter the spread of weeds and mean that the surface soil warms up more quickly in spring. Ideally they shouldn't be too big – a maximum of about 2 m (6 ft) wide – so that you need never walk over them. The soil structure is therefore retained, and all the beds will need is added organic matter and the occasional forking over to keep them in condition. Digging becomes a thing of the past with this method, which may appeal to many!

making a raised bed from gravel boards

For one raised bed you will need:
- 4 x gravel boards, 15 cm (6 in) deep, cut to the length specified on your plan
- 4 x posts made from 7.5 cm (3 in) square timber, cut to a length of 45 cm (18 in)

1. Work out from your design how much wood you will need to buy or get hold of. Treated gravel board or timber planks 15 cm (6 in) deep are recommended, but reclaimed timber, such as old scaffolding planks, can also be used.

2. Dig over and carefully level the area, using a spirit level. Remove any weeds. Mark out the beds with canes and twine.

3. Hammer in the posts with a sledgehammer, leaving 15 cm (6 in) above the level of the soil. Check with a spirit level that the posts are level at the top.

4. Position the first board against one of the corner posts and nail it to the post at the top and bottom, using 5 or 7.5 cm (2–3 in) galvanized nails.

5. Repeat with the remaining three boards. Firm down inside with a heap of topsoil.

6. Finally, fill the bed with topsoil and compost.

plan for a basic grid plot

2. ornamental plot

If you have time and are keen to make your plot aesthetically pleasing as well as productive, you could try this plan, which mixes flowers and vegetables in a decorative, potager-type plot. At the top of the plot, in front of the shed, is a seating area bordered by flowers, while opposite is a herb garden surrounded by low box hedges, divided into four diagonal sections. The central lawn circle is edged with lavender, but you could use rosemary or box instead. The beds are edged with flowers to create a wonderful splash of colour, as well as for the purposes of companion planting (see below), and the smaller beds around the circle are planted with decorative salad vegetables. Alternatively, these beds could be planted with flowers for an even more decorative feel. The bottom end of the allotment, with a fruit cage, compost area and cold frame, is screened from the rest of the plot by a row of attractive espaliered apple trees.

companion planting

Companion planting is the practice of growing certain plants together for the purpose of deterring pests and diseases. Although some people swear by it, results can be erratic, and it has never been scientifically proved as a method of pest control. Having said that, the general principle of biodiversity – that is, growing lots of different things together – can only be beneficial, and from an aesthetic point of view, growing flowers among your vegetables can look fantastic.

The theory behind companion planting is that with greater diversity and a mixed planting, insect pests find it harder to find their host plants (your precious crops) because they are distracted by different shapes, colours and scents. The disadvantage is that in a scheme like this the planting will be very dense, so you need to make sure that the soil is especially rich because there will be stiff competition for nutrients and water.

'Companion planting can mean a lot of different things to different people, but what it means to me is trying to make the plot more natural and attractive to predators of pests, as well as less attractive to the pests themselves. My aim is to create as natural an environment as possible in which my vegetables will grow. The flowers all attract welcome insects and help to keep my plot lively and more balanced in pest control than it would otherwise be if I just grew vegetables.'

Chris Smallbone, Manchester

Flowers such as these nasturtiums make good companion plants, attracting blackfly away from the vegetables and releasing a scent that is repellent to many pests.

good companion plants

There are plenty of unfounded theories on specific plant combinations, but I prefer to keep it general. The following flowering plants will be of general benefit on the vegetable patch.

• **French marigolds** (*Tagetes patula*) are the organic gardener's favourite flower. Marigolds repel many insects with their scent and can also reduce soil nematodes and slugs. They benefit a range of vegetables, especially tomatoes, and can look stunning on an allotment plot, bordering the beds or creating a boundary.

• **Nasturtiums** (*Tropaeolum* spp.) have bright flowers and scramble around the plot with ease, releasing a scent that is repellent to many insects. They also attract blackfly, so can be used as a 'trap crop' to lure the insects away from the vegetable.

• **Lavender** (*Lavandula* spp.), rosemary (*Rosmarinus* spp.) and other aromatics release chemicals that deter pests. Bordering a path with them is a good idea because every time you brush past them you'll release the scent.

• **Yarrow** (*Achillea millefolium*) attracts beneficial insects, such as hoverflies, which will help to control aphids.

• **Borage** (*Borago officinalis*) has pretty, star-like. blue flowers. The plant attracts blackfly away from crops such as broad beans.

• **Nicotiana sylvestris** has sticky stems that attract whitefly and thrips.

Other flowers that mix well with vegetables include cornflowers (*Centaurea cyanus*), catmint (*Nepeta* spp.) and chives (*Allium schoenoprasum*).

plan for an ornamental plot

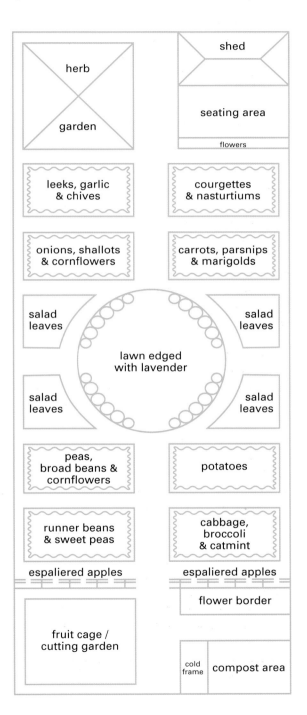

3. the low maintenance allotment

Is there such a thing as a low-maintenance allotment? Taking on an allotment means that you must be prepared to devote a certain amount of time to it, particularly in the spring and summer months – but there are time-saving devices that can help.

First, think carefully about what you are going to grow. Many crops need high levels of maintenance – especially those that are attractive to pests, such as the brassicas – but there are others that can be left for longer periods without attention. Consider planting a section of your plot with perennial fruit bushes, which need much less maintenance than annual vegetables. For example, autumn raspberries (which require even less maintenance than summer varieties) can be planted in a block to minimize weeds, as can gooseberries and blackcurrants. Globe artichokes (*Cynara scolymus*) are also perennial, producing spectacular silvery-grey foliage and edible flower buds, and their less edible cousin, the cardoon (*C. cardunculus*), is similarly decorative. Some people even grow vines on their allotments.

When it comes to annual vegetables, the mainstay of the plot, choose crops that take lots of space (to keep down weeds) but need minimal care. Potatoes are a good example, as are pumpkins, courgettes and squashes, which are relatively free from pests and diseases and can be left to their own devices (although they do get quite thirsty if the summer is hot and dry). Herbs are also easy, especially the Mediterranean sages and thymes, which need little water during summer.

Other ways to minimize the workload are to make raised beds (see page 36). Topsoil and compost are added to raised beds, so that the soil is never compacted, remains easily workable and will never need heavy digging. Mulching heavily can also help to conserve water during the summer months

Perennial crops such as artichokes or cardoons need less maintenance than annual vegetables.

(see page 57), although it won't, of course, totally eliminate the need to water.

Finally, perhaps you could consider giving over a section of your plot to ornamental plants (if the allotment rules permit it). Planting perennials or flowering shrubs will not only improve the look of your plot but keep weeds at bay too.

The plan for the low-maintenance allotment has a series of long, narrow raised beds, 1.4 m (4 ft 6 in) wide, which contain a range of easy-to-maintain vegetables. Wide paths are covered in bark chips to keep the weeds down, and the seating area in front of the shed is made from reclaimed brick. A trellis divides off the bottom third of the plot, behind which is a large fruit cage and two beds for larger crops, such as potatoes or pumpkins. These could also be used for large perennials, such as artichokes or cardoons, or for fruit trees.

tips for a low-maintenance plot

- choose crops such as potatoes that are low maintenance as well as being good ground cover

- plant perennial fruit bushes or easy ornamentals

- make raised beds to minimise digging

- cover soil when not in use to keep back weeds

- mulch regularly to suppress weeds and retain moisture

plants list

easy vegetables
- Beetroot
- Broad beans
- Climbing beans
- Courgettes and marrows
- Garlic
- Onions
- Potatoes
- Pumpkins and squashes
- Radishes
- Rocket
- Rhubarb
- Spinach

low-maintenance filler plants
- Blackcurrants
- Gooseberries
- Globe artichokes
- Raspberries
- Redcurrants
- Whitecurrants
- Vines

plan for a low-maintenance plot

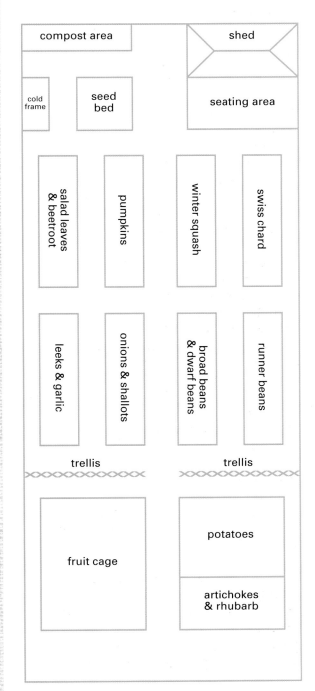

crop rotation

Crop rotation is simply the practice of growing vegetables in different parts of your plot from year to year in order to prevent the build up of soil-borne pests and diseases, such as clubroot (for brassicas) and white rot (for onions). If, for example, you grew potatoes in the same patch year after year, you would see a decline in the yield and health of the crop as the soil became more impoverished, while at the same time diseases specific to potatoes built up in the ground. However, while a degree of crop rotation is important, it isn't absolutely essential to follow a rigid scheme – it can simply be enough to bear in mind what has been grown in each bed the year before and make sure that the same crop isn't grown there for the next two or three years. However, well-organized gardeners can find it enormously helpful to have a good scheme to follow, knowing that it will result in enhanced yields and improve the overall health of the plot.

Three-, four- or five-year rotations can be followed, depending on how many different vegetables you are growing and how much space you have. Here we focus on a four-year rotation, with vegetable crops divided into four main categories. The fifth group 'others' can be planted with any of the four.

The first step in planning a crop rotation is to make a list of all the vegetables you want to grow, then group them into the categories shown opposite. There are three main groups – legumes, brassicas and root vegetables – plus the alliums (onions, shallots, leeks and garlic) and a group of 'others', which can be grown with any of the above to fill space. If you are planning a three-year rotation, the alliums can be grouped with the legumes, because they have similar needs from the soil. If, as shown, you are planning a four-year rotation, the alliums can be put into their own group.

Each of the main groups has different needs from the soil. Legumes require a reasonably rich soil, but more importantly, they are nitrogen 'fixers', meaning that they obtain nitrogen from the air and transfer it into the soil via their root systems. Brassicas are heavy feeders and need lots of nitrogen to thrive, so it makes sense for them to come after the legumes in the crop rotation – that is, they should be planted where the legumes were grown the previous year. Root vegetables don't need much nitrogen, so they can follow the brassicas. Alliums, too, don't make heavy demands on the soil, so they can follow the root crops, together with other vegetables, such as lettuces, sweetcorn and courgettes, which don't fit into any of the four main categories. These are classed as 'others' on the plan and can be slotted in wherever there are spaces.

The next step is to divide your plot up into sections. The long, narrow shape of an allotment lends itself to division lengthways, as shown on the plan, with different groups of crops at the top, in the middle and at the bottom of your plot, meaning that a rotation is easy to follow. But you can divide the space up in any way that you wish, depending on the size and shape of your plot, using any of the plans already described on pages 36–45 if you wish.

crop rotation groupings

alliums	roots	brassicas	legumes	others
Onions	Potatoes	Cabbage	Peas	Lettuce
Shallots	Carrots	Cauliflower	French beans	Sweetcorn
Garlic	Jerusalem	Brussels	Broad beans	Courgettes
Leeks	artichoke	sprouts	Runner beans	Pumpkins
	Parsnips	Broccoli		Squash
	Beetroot	Swede		Aubergines
		Turnip		Cucumbers
		Radishes		Chicory
		Oriental		Endive
		vegetables		Fennel
				Tomatoes
				Spinach
				Chard
				Celery
				Celeriac

four year rotation

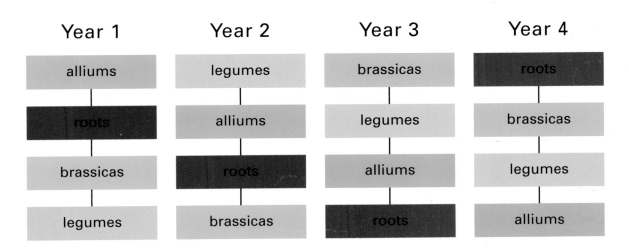

Year 1	Year 2	Year 3	Year 4
alliums	legumes	brassicas	roots
roots	alliums	legumes	brassicas
brassicas	roots	alliums	legumes
legumes	brassicas	roots	alliums

what to grow

This is the fun part – planning what you are going to grow over the next year and buying seeds. Obviously, what you grow is going to be determined by you and your family's preferences – what you like to eat, what you enjoy growing – but much of it is a matter of trial and error. A general rule is not to overplant. I found in my first year of allotment-growing that the quantities of vegetables I was growing were far too much for my family to get through, and some went to waste, although you'll find that there are always willing recipients of organic vegetables if you have a glut!

Think about what you would like to be eating through the year, and plant accordingly, making sure you have vegetables for each season, from spring cabbages to winter leeks. Often small quantities of crops can be sown in succession to prolong the harvest. If this is your first allotment year, I would say don't be too ambitious – keep to the basic vegetables and keep planting schemes simple.

F1 hybrids vs. heritage varieties

F1 hybrids are commercially produced seeds that have had unwanted characteristics removed from the plants' genetic make-up. F1 stands for 'first filial', meaning that it is the first generation of plants bred from a particular cross. These plants tend to be uniform in appearance and size, and they have been bred to be high yielding, reliable and disease resistant.

The disadvantage of F1 hybrids is that crops tend to mature all at once, which can be frustrating for small-scale gardeners. Seed from these plants should not be saved and replanted, because the second generation of plants will not come true. New seed must therefore be purchased the following year.

Heritage or heirloom vegetables

Heritage or heirloom vegetables, on the other hand, are old varieties that have been cultivated for many years. Often cropping over a longer period, they are more likely to grow well without chemical pesticides or fertilizers, but may be more variable in shape and colour. Many people swear that they taste much better than the commercial hybrids. Seed can be saved each year, with the plant retaining the same characteristics. However, unless you don't mind taking a chance you must take care to isolate wind- and insect-pollinated plants so they don't interbreed with similar plants (see page 89).

With the arrival of F1 hybrids, many old varieties are being lost, which is of great concern when the hybrids cannot be regenerated. To preserve the genetic diversity of these old varieties, seed exchange schemes have been set up in Britain, Europe and US. In Britain Garden Organic (previously HDRA, the Henry Doubleday Research Association) has the biggest seed 'library', where members swap seed of many old vegetable varieties.

preparing *the soil*

Understanding your soil and building it up by adding manure, compost and other natural substances are absolutely key to organic gardening, and before you even think about sowing your crops, you must carefully dig over and prepare the soil. There are many ways of improving the soil, and if you take time to nurture it and build it up, you will reap the rewards in the long term with healthier, stronger crops.

the science of soil

A basic understanding of how the soil works will help you get the most out of your plot – armed with this knowledge you are more likely to create the conditions that your plants need.

No matter what the soil type, about 90 per cent of its content is mineral (rock), and the remaining 10 per cent is made up with organic matter (humus), plus air and water. This living, organic matter is the crucial component of soil, and the more organic matter, the better quality the soil. There are billions upon billions of living organisms in the soil, from earthworms to minuscule bacteria, and they thrive on the organic matter that is added to the soil. During the process of decomposition the nutrients that are locked up in the organic matter are released into the soil and made available to the plants, including the three main nutrients that plants need for growth: nitrogen (N), phosphorus (P) and potassium (K) (see page 58).

soil pH

It is usually the mineral content of your soil that determines its acidity or alkalinity, and it varies widely throughout the country. It is measured on a scale of 1 to 14. The neutral point (and ideal for growing most vegetables) is 7, with 1–6 indicating an acid soil, and 8–14 alkaline. Acid soils are easier to correct, simply by incorporating lime into the soil (see page 77), but alkaline soils are more difficult to regulate. Simple pH testing kits are available from most garden centres.

finding out your soil type

One of the first things you should do when you inherit your new plot is get down on your hands and knees and feel your soil. Getting your hands dirty is a good way to discover what type of soil you have, and once you have worked this out you will know more about what its needs are. If the earth feels loose and light and runs easily through your fingers, then you have a sandy soil. If it is heavy and damp and easily rolls into a ball in your hand, then it has a high clay content.

All soils consist of organic matter plus three mineral elements – sand, clay and silt – in varying proportions. An easy way to see which mineral element makes up the highest proportion of your soil is to do this simple jam jar test. Half fill a jam jar with water, and then fill the remainder up with soil. Put the lid on and give it a good shake, then leave it to settle for a day or two. The sediments from the soil should have formed layers, with sand at the bottom (with the heaviest particles), silt in the middle and clay at the top, so you can see immediately which is in the highest proportion.

soil types

Loamy soil is the gardener's ideal. It is a well-balanced soil, high in organic matter – brown and crumbly in texture, high in nutrients and easy to dig. It is well drained but holds moisture well so it doesn't get too dry in summer. A good loamy soil is what we are all aiming for, and it is perfectly achievable to create such soil by persistently adding organic matter over a long period.

Clay soil is very common and can be hard work initially. Made up of small soil particles, it is heavy and sticky and often hard to dig. In winter it can get waterlogged, while in summer it can bake to a hard, dense crust, often with surface cracks, excluding air from the soil and making it difficult to get water to the roots of the plants. The advantage of a clay soil is that it is high in nutrients, so once it has been made more workable it can produce very good results.

Sandy soil feels loose and gritty when handled and is easy to dig over. Made up of large soil particles, it is free-draining, so won't get water-logged in winter, but does get dry in summer. It isn't a rich soil because its loose texture means that nutrients are washed away easily with rain.

Chalky soil is easy to identify by its white clumps of chalk or flint. It is a shallow soil and naturally free-draining, which means that it can be very dry in summer. It is also alkaline, so probably won't need additional lime.

Silty soil is made up of fine grains, originally the deposit from a river. Like sandy soil, it is free-draining and easy to dig but can be low in nutrients. It is also easily compacted underfoot and after heavy rains.

Peaty soils are less common, but where they do occur – in the Fens of East Anglia, for example – they make some of the county's best farmland. The soil is rich, black and fertile (although acid), with a sponge-like texture that locks in nutrients.

improving your soil

With that crumbly, dark, nutrient-rich loam as your goal, there is one way to improve any type of soil and that is to add organic matter, either by digging it in or by mulching. In lighter, free-draining soils, this organic matter will improve water-retention and add nutrients, while in heavier soils it will open up the soil structure and improve drainage.

When you start out, you probably won't have a store of compost on tap to add to the soil, so either get hold of a large load of manure to dig in or invest in a dozen bags of good organic soil improver just to get things going.

There are many different types of composts on the market, from spent mushroom waste to composts made from bracken and leaves. For clay soils, where bulk is most important, a low-nutrient compost can be used to improve the soil structure. For lighter, poorer soils a nutrient-rich manure is best. Most crops, apart from root vegetables such as carrots and parsnips, will benefit from an annual dressing of manure. It is best to apply it in autumn, either by leaving it on the surface of the soil for the worms to do the work or by digging it in.

to dig or not to dig?

There is great debate nowadays as to whether digging, and particularly the old-fashioned double-digging method, is absolutely necessary. Often heavy-duty digging is necessary only as an initial measure when you are clearing your plot, but it also depends to a certain extent on what kind of soil you are dealing with.

On heavy clay soils digging is necessary to break up the soil and make it workable, but on sandy or silty soils it is not essential, because the texture of the soil is already loose. In these instances, the best approach is a light forking over in spring followed by the addition of a thick mulch. On heavier clay soils it is best to dig in autumn, before the first frosts, and when the soil is neither too wet nor too dry. Turn over the soil to bury any annual weeds but leave the clods intact for the frosts to break down.

double digging

A traditional method, double digging means cultivating the soil to a depth of two spade heads. It is time-consuming and hard physical work, but can give good results, especially when cultivating a new plot or if the soil is heavy and compacted.

how to double dig

1. Dig a trench about 30 cm (12 in) wide, and the length of the area you want to tackle. Dig out the topsoil from this trench to the depth of one spade and pile it to one side.

2. Add a layer of animal manure in the bottom of the trench and, using a fork, dig over to the depth of the fork. Loosen the subsoil and incorporate the manure.

3. Dig another trench parallel to the first, but this time throw the topsoil into the first trench as you go. Then repeat step 2.

4. Continue digging trenches and repeating the process until you have covered the planned area. When you have dug the final trench, fill it up with the topsoil that you put aside from the first trench.

Depending on your soil type, digging may not be necessary at all, especially if you mulch regularly.

mulching

A mulch is simply a layer of organic or inorganic material that is laid on top of the soil to suppress weeds and conserve moisture by preventing water evaporation. As a rule, the mulch needs to be at least 5 cm (2 in) thick to work adequately. If the mulch is organic, it doubles up as a soil improver by leaching nutrients down into the soil and improving the soil structure as earthworms gradually pull it down. The benefits of mulching cannot be over-stated, and it is almost essential in vegetable gardening, particularly on an allotment plot, which may not be visited every day. Nothing can beat homemade compost as a good all-round mulch, but there are a number of other types of organic mulch that can also be used.

organic mulches

Garden compost is probably the best all-round mulch and is readily available if you are making your own.

Farmyard manure, in the form of horse or chicken manure mixed with straw or bedding material, is the most popular. Both types are full of nutrients but must be well-rotted before being used because the ammonia content in unrotted manure may scorch and damage plants.

Grass clippings are not so high in nutrients but are good for suppressing weeds and keeping moisture in.

Leafmould, made from well-rotted leaves, is an excellent mulch, low in nutrients but excellent for improving soil structure.

Mushroom compost, the after-product of mushroom farming, is high in nutrients but slightly limed, so it is best not to use it on a very alkaline soil.

Cocoa shells make a slow-rotting, nutrient-rich mulch that forms a dense layer that s good for both weed suppressing and moisture retention. Care must be taken because it can be poisonous to dogs.

making compost

Of course, once your allotment is up and running it makes sense to make your own compost – in fact, it is almost a crime not to! Composting is a way of recycling all the vegetable and weed waste from your plot and putting back into the soil what you have taken out. It is nature's best soil conditioner.

step one in making your own compost is to make or get hold of a good compost bin. These range from plastic bins, cheaply available from most borough councils, to homemade bins made from wire or wooden slats. You will need at least two compost bins on a full-size allotment because you will be generating large quantities of waste.

the key to making good compost is to ensure that there is a good balance of materials going into the heap, as well as adequate supplies of air and water. The best and swiftest decomposition takes place when there is an equal mix between nitrogen-rich and carbon-rich matter. Looking at it in a more visual way, the things that are high in nitrogen are soft and green (kitchen waste, grass clippings), while the carbon-rich materials are hard, woody and brown (leaves, straw, woody waste from shrubs). Adding these materials in layers can speed up decomposition, but the most important thing is to ensure that they are well mixed together and ideally shredded or chopped so that more surface area is exposed to the organisms active in the decomposition process.

air and water are the other crucial components, and decomposition will slow down if either is in short supply. Once these four elements – carbon, nitrogen, air and water – are well-balanced, the compost heap will take off, rotting swiftly down and producing heat in the process. Turning the heap regularly will speed things up even more by injecting more oxygen into the heap as everything rots down.

People sometimes wonder why their compost takes three years to rot down when all they are adding is grass clippings and kitchen waste. Both these substances are high in nitrogen and tend to form soggy layers that exclude air. Grass clippings in particular can form a thick mat that won't allow air to penetrate, preventing normal aerobic decomposition. But don't be put off. If you mix grass clippings with woody, brown material, such as straw or hedge clippings, the air will be able to penetrate and everything will rot down much more happily together.

In summer, when the weather is warmer, you can have useable compost within a month or two of starting off – but only if you spend time building, mixing and turning the heap. More realistically, you will probably be looking at four or five months before harvesting your first compost.

compost accelerators

There are ways to speed up the decomposition process for those who suffer from impatience. You can buy proprietary compost accelerators, which contain millions of dormant soil organisms that come to life when they come into contact with water. Other accelerators include comfrey and nettles, both high in nitrogen, which can give a boost to a slow compost heap. And finally, that substance that all of us have on tap – urine!

ingredients for the compost heap

nitrogen-rich materials (green)

Animal manure
Annual weeds
Comfrey
Grass clippings
Kitchen scraps
Nettles
Soft prunings

carbon-rich materials (brown)

Bark
Cardboard
Dead leaves
Newspaper
Sawdust
Straw
Tree prunings
Wood chips
Woody plant remains

don't compost

Animal faeces
Meat or dairy products

recycling tip

Make compost bins from old wooden pallets:
You'll need five pallets – one for the base, and
four for the sides. Simply wire three sides
together and attach them to the base, and then
attach the fourth so that it can be opened and
closed to allow you to remove the compost.

'The best way to rot down
garden rubbish into compost is
to pack it into a pit dug in an
odd corner or to stack it up in a
heap. As each layer is put down
it should be dressed freely with
slaked lime or new soot.'
The Smallholder Encyclopaedia, 1937

making leafmould

Leafmould, a compost made from leaves alone, is an invaluable soil conditioner. If you have lots of trees around your allotment or surrounding your garden at home, it might be better to make leafmould separately, rather than adding the leaves to your compost heap, because leaves are extremely high in carbon and can take a long time to rot down. Although low in nutrients, leafmould has a light, crumbly texture and is very good for improving soil structure. It can also be used for making your own potting compost.

To make leafmould, simply gather together fallen leaves and pile them into a wire mesh enclosure. Alternatively, put them into black plastic bin liners with several holes pierced into the bottom and sides to allow aeration and drainage. Depending on the leaf variety it may take two or more years for the leaves to rot down, but decomposition can be accelerated by shredding the leaves before piling them together. If you don't have a shredder, just run over them with a lawn-mower. The only other thing to remember is to keep the leaves slightly damp.

using green manures

Green manures are usually leguminous plants that are sown and grown for a season, before being dug back into the soil as a source of nutrients for the plants that go into that piece of ground. Sowing a green manure crop is advantageous in several ways, particularly on an allotment plot where you are gardening on potentially quite a large space.

First, the green manure acts as a groundcover, suppressing weeds and protecting the soil from erosion and moisture loss. Second, the plants have deep root systems, which break up the earth and improve soil structure. Most importantly, they actively feed the soil. Leguminous plants (members of the pea family) are unique in that they are able to 'fix' nitrogen from the air and transfer it to the soil via little nodules on their roots. In addition, the young growth is high in nitrogen, so that, too, enriches the soil when it is cut back and dug back in.

Green manures are extremely easy to grow. Seed is broadcast over a prepared bed and then gently raked in. The plants are left until they reach a certain stage of growth (usually just before they flower so that they don't set seed and start popping up everywhere else the following season) and then cut back and dug into the soil to rot down. Some varieties are suitable for late-autumn planting, to be dug back in before sowing vegetable crops in spring, while other quick-growing varieties can be sown in early spring and dug back in during the summer.

It is advisable to wait several weeks before replanting, because the high nitrogen levels in the soil may be detrimental to young plants. The only other caution with green manures is not to let them grow for too long because they can become tough and consequently take much longer to rot down.

A traditional bonfire bin could make an excellent container for making leafmould.

green manures

for overwintering

Trefoil, black medick (*Medicago lupulina*) is easy to grow; sow from mid-spring to late summer.

Alfalfa, lucerne (*Medicago sativa*) is deep-rooted and good for soil structure; sow from mid-spring to midsummer.

Phacelia (*Phacelia tanacetifolia*) has ferny foliage and blue flowers; sow from mid-spring to early autumn.

Grazing rye (*Secale cereale*) has an extensive root system good for soil structure; sow from late summer to late autumn.

Winter tares (*Vicia sativa*) is a very swift-growing nitrogen-fixer; sow from early spring to early autumn.

for spring or summer cropping

Buckwheat (*Fagopyrum esculentum*) is an attractive, deep-rooted plant; sow from early spring to midsummer.

Bitter lupin (*Lupinus angustifolius*) is a very good nitrogen fixer; sow from early spring to early summer.

Crimson clover (*Trifolium incarnatum*) is an attractive plant that is vigorous and swift-growing; the roots do not penetrate deeply; sow from early spring to midsummer.

Fenugreek (*Trigonella foenum-graecum*) is swift growing with lots of foliage; sow from early spring to midsummer.

Planted in spring or early summer, fenugreek is one of the fastest-growing green manures, with masses of foliage to provide good ground cover.

'I use a mixture of oats and broad beans for my green manure crops – oats because I use them for feeding the chickens and always have them handy, and broad beans because I save them myself, so they cost nothing! The main point is: the nutrients in the soil get locked up in the green manure crop and they're then safe for the winter – the rain can't leach them out of the plants as it can out of the soil. Then when you dig them in, they're released again.' *Ben Kissane, Co. Kerry*

sowing and growing

Growing plants from seed is one of the most satisfying and rewarding things you can do, and it is not at all difficult provided you follow a few basic techniques. Most vegetables are grown annually from seed, and are sown either direct into the ground or inside, to be pricked out and later transplanted into the ground.

sowing inside

If you are lucky enough to have a greenhouse, you can give some crops a head start by sowing them in seed trays or pots early on in the year. It is preferable to start some vegetables – tomatoes and courgettes, for example – off indoors, and they must be hardened off sufficiently before planting out (see page 67) because they are too tender to be planted direct into the ground. Other less tender crops respond well to this treatment too, the advantage being that they are bigger and hardier when planted out, offering more competition to weeds and more resistance to pests and disease.

The greenhouse must be insulated and heated to prevent the temperature falling below freezing, and ideally keeping it at 3–5°C (37–41°F). If you don't have a greenhouse, seeds of some tender plants, such as pumpkins and squashes, can be started off on windowsills indoors.

heated propagators

Propagators are useful but by no means essential. Usually consisting of a seed tray placed on top of a heating element, a propagator provides gentle heat from the bottom, which will raise the soil temperature by about 10°C (about 18°F). Seeds are sown into the tray, and then a lid is placed on top to keep the air moist.

sowing in pots or seed trays

Fill clean pots or seed trays with multipurpose or seed compost and firm it down gently so that the surface is several centimetres below the rim of the pot. Water the compost using a fine rose on a watering can so that it is damp. Sow small seed on the surface of the compost by scattering it lightly all over using your fingertips, and cover lightly with compost or vermiculite. Finally, place a layer of clingfilm, sheet of glass or a plastic sandwich bag secured with an elastic band over the pot to prevent the compost from drying out. The compost should be kept damp, but not too wet, which can cause damping off, a fungal disease that is caused by overwatering, so turn the glass or change the clingfilm regularly to minimize condensation. Sow larger seeds, such as pumpkin or courgette, two to a pot, pushing each just under the surface of the compost. The seedlings should be kept at a warm, uniform temperature until they have emerged; after this point a cooler temperature is preferable. Too much heat can make the seedlings leggy and weak.

sowing in modular trays

Sowing in modules prevents root disturbance caused by pricking out or transplanting, and can be easier, especially if planting into biodegradable modules or newspaper pots, which are left in the soil. Scatter small seed thinly on the surface of the compost in a module, and remove weaker seedlings gradually until a single, strong specimen is left. With larger seed, sow two or three seeds to a module, and remove the weaker seedlings. Modules can also be used at the second stage, for pricking out.

Sowing seedlings into modular trays prevents the root disturbance that occurs when pricking out or transplanting, as well as saving time.

recycling tip

Make biodegradable seed pots from newspaper:

1. Open a full sheet of newspaper and lie it flat.

2. Fold it in half lengthways and then again, so you have a long, narrow strip of folded newspaper.

3. Roll the newspaper around a tall, straight-sided glass, with about 20 cm (8 in) of newspaper above the open end of the glass.

4. Push the ends of the newspaper into the open end of the glass.

5. Remove the glass so that you have the newspaper pot in your hand.

6. Push the bottom of the glass into the newspaper pot, squashing the folded bottom to flatten it, and then remove the glass. Once the pot has been filled with soil, it will be secure.

7. Fill with seed compost and stack the seed pots side by side in a plastic tray to support each other.

8. Sow chosen seeds and wait for them to grow. Once they are big enough, they can be planted out, pot and all. The newspaper will break down quickly, allowing the roots to spread.

pricking out

If small seeds have been sown all together in a container, they will need pricking out (separating and replanting). This is usually done when they have developed two pairs of leaves and are big enough to handle. If seedlings aren't pricked out they will grow weak and leggy, and probably collapse.

Before pricking out, prepare new seed trays or pots, filled with multipurpose compost and firmed down as before. Water the seedlings thoroughly. Loosen the seedlings by using a lollipop stick or plastic plant label, and gently pull them out with your forefinger and thumb. Then, using your forefinger or a dibber, make holes in the compost and place a seedling into each hole, spacing them about 10 cm (4 in) apart and planting to the same depth as before, with the first set of leaves sitting on top of the compost. The final steps are to firm down the compost gently around the seedling with your fingertips and then to water them in, using a fine rose on a watering can.

tips for sowing and growing indoors

- sow seedlings at the recommended temperature and keep at an even heat until they have emerged.

- Keep seedlings in full light, and turn the trays regularly to stop the seedlings bending over.

- Don't wait too long before pricking out the seedlings or they'll become weak and leggy.

- Sow in modules to save time and prevent root disturbance.

- When pricking out, take care to hold seedlings by their leaves, not the stems.

- Harden off all your seedlings by gradually acclimatizing them to outdoor temperatures.

hardening off

All seedlings raised inside need hardening off –
that is, gradually acclimatizing to temperatures and
winds outside to make them stronger. Hardening
off is usually done in late spring and takes several
weeks, depending on how tender the plants are
and how warm they have been since germination.
Plants raised on a warm windowsill or heated
propagator usually take the longest as they have
soft new growth.

Start by taking the seed trays outside on a fairly
warm day, remembering to take them in again at
night. Alternatively, if you have one, a cold frame
is ideal, because it can be fully closed to start
with, keeping temperatures up and wind out, with
ventilation gradually increased as time goes on,
and the cover eventually left permanently open.

transplanting seedlings

You can plant out your seedlings once they have
been hardened off for between two and six weeks
and are sturdy enough to withstand the open air.
This is usually at a time when the risk of frost has
passed. Seedlings should be watered thoroughly
before and after transplanting.

sowing outside

The most common and easiest way of sowing vegetable seed is directly into the ground outdoors, and most seed can be sown in this way. The only disadvantage with this method is that you have to wait until the soil temperatures are high enough, which may delay your harvest. Some crops, such as brassicas and leeks are sown in seedbeds and later transplanted to their final growing spot, while others, such as peas and carrots, are left to grow in the same position they were sown in.

The question of when to sow the seed varies from crop to crop (see Vegetable Directory, page 94), and usually depends on soil temperature and hardiness. Hardy vegetables, such as broad beans, can be sown in early spring when the soil temperature is still low, but others are more susceptible to frost and must be planted after the last frosts. Soil thermometers are widely available and easy to use.

Above and opposite: Most crops can be planted directly outside in drills that are marked out with a garden line and made with the edge of a hoe or a trowel.

planting in drills

Before you sow any seed you must prepare the ground thoroughly so that the tiny germinating seedlings don't have to battle to break through the earth. Dig or fork over the bed, removing any weeds or stones as you go. Rake the area to a fine tilth, firming it down with your feet and using the back of the rake to break up any clods of earth. Most seed is planted in long, narrow drills, which can be marked out with stake and twine or a garden line.

Using the corner of a hoe or a trowel make a shallow drill about 2.5 cm (1 in) deep for small seeds (deeper for larger seeds, such as beans, sweetcorn and peas). If the soil is very dry, it is sensible to water before sowing. Scatter small seed as thinly as possible along the drill, and then cover with a thin layer of topsoil before watering with the fine rose on the watering can. Larger seed can be sown in stations along the drill, sowing two or three seeds into holes made with a dibber or your finger.

Some vegetables, such as peas and radishes, are best sown in wide drills. These are made with a wide draw hoe or spade and should measure about 20 cm (8 in) across. Space the seeds evenly across the bottom of the drill.

'In the case of vegetables which can be transplanted successfully, e.g., beans and onions, it is a good plan to sow a small clump at the end of the row from which gaps can be repaired.'
The Vegetable Garden Displayed, 1942

broadcasting seed

Broadcasting seed is a method not as commonly used as drills, but it is useful for planting blocks of plants such as cut-and-come again salad leaves or a patch of rocket. Prepare the seedbed as for drill and then simply scatter the seed as evenly as possible over the area. Rake it gently in to pull the soil over the seed, then water thoroughly.

thinning out

Thinning out is the removal of surplus seedlings in order to let the chosen few perform as well as possible – and sowing as thinly as possible in the first instance will help make this task less arduous. As soon as the seedlings appear, the process can begin – nipping out the weaker plants and leaving the bigger ones – and over the course of two or three weeks, the line of seedlings will gradually be reduced to their final spacings.

tips for sowing outdoors

- Make sure the soil is warm enough (at least 10 degrees) before sowing hardy vegetables, and wait until all threat of frost has passed before sowing more tender crops. Remove stones and rake thoroughly.

- Don't sow seed too deeply, or it will struggle to germinate. Small seed eg. carrots should be covered lightly to a depth of 1–2cm

- Sow as thinly as possible. Use thumb and fingers rather than shaking seed directly from the seed packet.

- Where dry, water seed every other day, with the fine rose on the watering can.

- Thin seedlings soon after they appear.

transplanting

Some plants, such as leeks and some members of the brassica family, should be sown initially into a prepared nursery bed – that is, a small seedbed where they can be planted in short rows – and later transplanted to their final positions. This is primarily because they need lots of space to mature, so it is a waste of space to grow them at their final spacings when they are small. As a rule, transplant brassicas when the seedlings reach about 15 cm (6 in) high. Dig them up with a trowel, taking care with the roots and leaving as much soil around them as possible. Use a trowel to dig holes large enough to accommodate the rootball, and drop the seedlings in so that the bottom pair of leaves sit on the surface of the soil. Fill in with soil, and firm around them with the heel of your hand. When you are transplanting leeks, wait until they have grown to the width of a pencil. Make deep holes with a dibber and drop the seedlings in.

Cabbages (top) and leeks (bottom) are always started off in a seed bed before being transplanted to their final growing spaces.

vegetative propagation

taking cuttings

Perennial fruit bushes, such as currants and gooseberries, can be propagated by taking hardwood cuttings at the end of the growing season. You will need space for a nursery bed in which to plant these cuttings, where they will be left for a year. Prepare a narrow trench, about 15 cm (6 in) deep, with a layer of grit or gravel in the bottom. Cut 30 cm (12 in) lengths of vigorous, well-ripened stem, the lower cut going straight across and the top cut slightly slanted (so that water doesn't accumulate). Remove all but the top three or four buds, and then insert the cuttings into the prepared trench, about 15 cm (6 in) apart. The cuttings will root slowly over the course of a year and won't need much looking after apart from watering if the weather is very dry. The following autumn, they can be lifted and transplanted to their new growing position.

dividing rootstock

Rhubarb can get woody and congested if it isn't divided every four or five years, resulting in weak growth and poor-tasting stems. The crowns can easily be divided during the plant's dormancy in late autumn or early winter. Lift them with a fork to avoid damaging the roots, and then split the crowns into sections with a spade, making sure that each one has a good thick section of root and at least one strong bud. Discard old, woody material and replant the smaller sections into soil prepared with lots of well-rotted manure. Plant them 1 m (3 ft) apart, with the buds just below, or at, soil level.

Globe artichokes and cardoons can be propagated by the offsets that they produce around the crown. Use a sharp-edged spade to sever the offsets from the crown, and replant them in fresh soil in late spring.

layering

Blackberries, loganberries and other hybrid fruit can be increased by tip layering, a simple method that involves burying the tip of a shoot to encourage further rooting. In spring choose a long, arching stem that will reach the ground from the parent plant. Simply bend it over and bury the tip under the soil, pegging it down with a loop of wire if the soil is not heavy enough to hold it. New roots will develop and a new plantlet will grow. Usually by autumn, the new plant should be ready to grow on its own, and the stem from the parent plant can be severed.

lifting suckers

Strawberries increase by sending out suckering runners that are easy to propagate from. If you have just planted your strawberry bed, it is best to wait two or three years before propagating from the plants, so that they can increase their fruit-bearing vigour. Nip off the runners as they are produced over the course of the summer. After the first couple of years you can use the runners to produce new plants. During the summer months, peg down runners into 9 cm (23 in) pots half-buried into the soil. New plants will soon root, and in autumn these can be planted in a fresh, well-prepared bed. Old plants should be destroyed.

other growing methods

catch cropping

If you are sharing an allotment or have a half- or quarter-size plot you will need to think about ways of using space efficiently. Catch crops – small, swift-growing crops that can be planted between rows of other vegetables – are useful for maximizing space. The idea is that they grow and mature quickly and are therefore harvested before the main crops around them get too big. Ideal candidates for catch cropping are radishes, beetroot, spinach, as well as lettuces and other salad crops.

'Lettuces and radishes above all other vegetables should be sown repeatedly in short rows in order to provide a continuous succession of young material to be gathered as it can be used.'
The Vegetable Garden Displayed, 1942

types of catch crop

- Plant rows of **radishes**, which mature in 4–8 weeks, between slow root crops like parsnips.

- Plant **lettuces**, which mature in 10–12 weeks, between rows of runner beans before they get big enough to cast shade.

- Try **turnips**, which mature in 6–10 weeks, between slow-maturing brassicas.

Swift-growing lettuces and other salad vegetables make excellent catch crops, planted in rows between other slower maturing crops.

double-cropping

To maximize space, once an early crop has been harvested, it can be followed by another sowing so that the beds are not left empty. For example early potatoes can be followed by tender, late-starting crops, such as courgettes and squashes, that were started off in pots. Other late-season vegetables, eg., leeks and winter brassicas, can be planted or transplanted as beds become available.

getting the most from your plants

Having coaxed your seeds into life, they will need a good deal of nurturing over the coming months if you are to get the best from them. Different crops have different needs. Some are less demanding and may need only minimal watering, but others require feeding or protecting to produce a good harvest. At the very least, regular weeding is essential to make sure there isn't competition for precious nutrients. If you put in the time at this stage, you'll be rewarded with healthy crops and bumper harvests.

feeding your crops

All plants require three main nutrients for growth – nitrogen (N), phosphorus (P) and potassium (K) – as well as a host of other trace elements, which are present in the soil in smaller quantities.

Nitrogen is essential for the formation of plant tissue and leafy growth. Demand for it is particularly high when the plant is young and growth is rapid. Signs of nitrogen deficiency are poor growth and pale green, sickly-looking leaves. Phosphorus is needed for root development, and it helps the plant to retain water. Signs of phosphorus deficiency are red or purplish discolorations underneath the leaves. Potassium is used most notably in the development of flowers and fruit. Signs of deficiency include brown, curling leaf edges and lack of flowers and fruit.

All three nutrients are provided by soil organisms in the process of decomposing organic matter – so you are off to a good start if you have already prepared the soil with lots of manure or garden compost. If your soil is good, having been fed and nurtured for a number of years, then fertilizers in any form really aren't necessary – in fact Garden Organic, Britain's leading organic body, goes as far as to say that fertilizers should never be used in good organic practice. But if you are just starting out on your allotment, some of your crops may need an extra boost in spring to maximize your harvests. Brassicas, generally, are greedy feeders and will benefit from a nitrogen-rich fertilizer, while a phosphorus-rich fertilizer, like bonemeal, can benefit potatoes and other root crops. Organic, non-chemical fertilizers made from plant or animal sources are widely available to buy, but you can also make your own liquid fertilizers from manure, compost or plants such as comfrey or nettles.

organic vs. chemicals

Chemical fertilizers are widely available but are used less and less nowadays. They provide large injections of nutrients and work very fast, promoting unnaturally swift growth. This can sometimes take the flavour away from vegetables, making them watery, and also opening them up to attacks from pests and diseases. Perhaps more than anything else though, the chemicals in these fertilizers damage the delicate balance of the soil, destroying the micro-organisms that keep the soil healthy, so you are actually doing more long-term damage than good by applying them.

organic fertilizers

Bonemeal is a phosphate-rich fertilizer useful for promoting strong root growth.

Chicken manure pellets are high in nitrogen and good for leafy vegetables.

Fish, blood and bone is a good balanced all-round fertilizer.

Hoof and horn is nitrogen-rich and good for leafy vegetables.

Rock potash is high in potassium and is good for potatoes and tomatoes.

Seaweed meal is a general tonic with many trace elements.

Wood ash from your own bonfires is high in potassium. (Don't use coal ash because this contains toxins and is lower in potassium.)

liming the soil

Many people talk about adding lime to the soil, and it has been widely practised for many years, yet the reasons for doing it are often not fully understood. Adding lime to the soil helps to get the balance between the soil's acidity and alkalinity, measured on the pH scale ranging from 1 (highly acid) to 14 (highly alkaline). A pH of 7 is neutral and the optimum for most vegetables. In soils with a pH outside this range plants may show signs of nutrient deficiency. In acid soils nitrogen, potassium and magnesium all become less readily available to plants, while in alkaline soils phosphorus is reduced. Adding lime to an acidic soil can help to restore this balance, while at the same time providing an important source of calcium.

Liming can also indirectly help the soil structure. By improving root development and raising the numbers of micro-organisms, the crumb structure of the soil is improved, therefore improving the performance of your plants. A final plus-point for lime is that it can help prevent diseases such as clubroot in the brassicas, and because brassicas generally prefer slightly alkaline conditions, liming these areas can be beneficial.

However, you should lime your soil only if its pH is below 6.5. If you do need it, it should be applied in autumn, just before digging the soil over. It should not be applied at the same time as you add organic matter or fertilizers.

liquid manures

Liquid manures, sometimes called tonics or 'teas', are easy to make and, because they are made available immediately to the plants they are excellent for a quick-fix burst of nutrients. They can be applied directly on to the leaves (known as a foliar feed) and are instantly absorbed into the plant's infrastructure. The disadvantage of these feeds, however, is that they are used up immediately and don't remain in the soil, so ideally they should be used regularly (every two or three weeks) for best effect.

They are made simply by steeping manure, compost or nutrient-rich plants in water, putting them inside a hessian sack or an old pillow-case. The liquid should be diluted with four parts water before using so that it does not scorch the plant's leaves.

comfrey and nettles

Comfrey is one of the wonder-plants of the allotment. It is a swift-growing, tough plant, which can be grown either from seed or from a section of root.

The large, nutrient-rich leaves (high in potassium and nitrogen) can be shredded and scattered directly on to the ground as a mulch, added to the compost heap as an accelerator or steeped in water to make a nutritious liquid manure or tea.

Nettles are especially high in nitrogen and can be treated in a similar way to comfrey to provide a liquid manure.

comfrey tea

To make comfrey tea, shred a large armful of leaves and put them in a large bucket. Fill to the top with water and leave to soak for two or three weeks. The resulting dark brown liquid smells revolting, but it makes a wonderful tonic for plants that need a potassium and nitrogen boost. It can be very strong so must be diluted to the colour of weak tea. Members of the Solanaceae family (tomatoes and potatoes) that need extra potash will be especially grateful for a dose of comfrey tea.

watering

Watering, of course, is essential over the long summer months, but your aim should be to find ways to minimize watering at every step. The water-retentiveness of the soil is directly related to the amount of organic matter in it. Humus is like a sponge, soaking up the available moisture and holding it in the soil to be made available to the roots of your crops. So if your soil is too light, with a low percentage of organic matter, water will drain away too quickly, and you will find yourself having to water more frequently.

If you have clay soil, you will need to water less often, but without enough organic matter, the problem is the soil will become waterlogged during periods of heavy rain, and this presents its own problems for your crops. Regular thick mulching with well-rotted manure or another source of organic matter is the key to conserving moisture during the summer months.

Different allotments have different rules when it comes to water supply and demand. Some have good networks of pipes and taps, so that a tap is never far away. Others have only a single water source, which plot-holders have to share. Most forbid the use of hosepipes, so there is much to-ing and fro-ing with watering cans.

On my former allotment, water was pumped up from the water table via a dozen archaic pumps (I was lucky enough to inherit a plot with a pump), and I know other allotmenteers who have made their own wells. This is a practice that should be researched thoroughly before you excavate a well because there are obvious dangers associated with it. It can be done only where the water table is not too deep (on clay soils rather than sandy) and when the allotment association gives permission.

Whatever provision for water there is on your allotment, it makes sense to start your own source. Water butts can be positioned next to sheds with guttering and downpipes to save rainwater. If you don't have a shed, you can improvise by using tarpaulin or corrugated iron to channel the water into a container, as shown, left.

water-wise tips

- **Mulch regularly** with organic matter to retain the moisture content of the soil.

- **Think about** the vegetables you are watering and **their needs**.

- **Sink a funnel, pipe or plastic water bottle** into the ground next to a plant to direct water to the roots (this is a good tip for courgettes or squashes).

- **Draw up a ridge of soil** around widely spaced crops to make a well of water (again, this works well with courgettes or squashes).

- Where allowed, **use a leaky hose system** to water your thirstiest crops; this is more efficient than a sprinkler because water is not lost in the air.

- **Water in the evening** so that less is lost to evaporation; watering in the middle of the day when it is sunny can scorch the plants.

what needs watering?

Some vegetable plants need large amounts of water, while others need very little, and it is worth thinking about this before dousing the whole of your allotment. Whatever you do, remember the crucial times for watering are when sowing and transplanting, when the roots are forming or at their most vulnerable.

Once growing, the plants most in need of water are the leafy vegetables, such as spinach, lettuces and cabbages, which will need regular watering at least once a week, and more often in hot weather. Potatoes, courgettes, squashes, leeks, radishes and runner beans also need regular watering.

Other crops need watering only when the fruit has started forming – for example, French beans, peas, broad beans and sweetcorn – and too much water before this point will promote foliage at the expense of yield.

Root vegetables and onions are much less thirsty than other vegetables, and once they are established they can do without watering in all but the hottest of weather.

Pumpkins and squash need regular watering in dry weather.

how much water?

thirsty vegetables

Aubergines, cabbages, cauliflowers, courgettes, cucumbers, leeks, lettuces, oriental greens, peppers, early potatoes, pumpkins and squashes, radishes, runner beans, spinach, tomatoes

thirsty only when setting flower or fruit

Broad beans, broccoli, Brussels sprouts, French beans, peas, sweetcorn

not thirsty at all

Beetroot, carrots, garlic, onions, parsnips, shallots, swede, turnips

Onions and shallots are some of the least thirsty vegetables on the plot.

weeding

We looked at ways to control perennial weeds in Chapter 2, and now we come to annual weeds. These are much less of a problem, but nevertheless must be kept on top of because they can swiftly take over an allotment plot, particularly if you let them get to the flowering and seed-setting stage.

Even in their infancy, weeds provide competition for water and nutrients in the soil, so they must be removed if you are to get the best yields from your vegetables. Hoeing is the best method of elimination, and if it is done regularly and the weeds are very small, it really isn't an onerous task – cutting the tops off annual weeds is sufficient. In areas where hoeing is difficult, hand-pulling the weeds will have to do. Mulching once again is the best option, because it will suppress the weeds.

Use a hoe (below) to control annual weeds. Perennials (above) require more thorough removal of all parts including roots.

protecting crops

Protecting your crops from the weather or wildlife is often necessary, and there are a number of ways you can do this, from using cloches and cold frames to covering an area with netting.

frost

Cold frames are extremely useful on an allotment and can often be made from reclaimed materials, such as old windows. They are used for hardening off seedlings, as well as for growing tender vegetables such as aubergines or melons. Cold frames with glass sides maximize the light for your plants, but are cooler than those with brick or wooden sides.

Cloches are essentially movable 'houses' for individual plants or groups of plants, made from glass or plastic. They are designed to protect plants from the cold and the wind as well as giving some defence against pests. They are especially useful in early spring when the weather is still unpredictable and mean that you can plant crops earlier than you would normally do. They can also be placed over an area in which you are planning to sow to warm up the soil by a few degrees.

Made from glass or plastic, cloches come in a variety of forms, from individual glass domes to a glass 'tent' system, where sheets of glass are clipped together at the top to form a triangle, and then placed side by side to make a tunnel. Mini-cloches can be made by cutting large plastic bottles in half, and these are excellent for small seedlings that need a helping hand. As the weather gets warmer, you will need to acclimatize crops grown in this way by slowly ventilating the sides of the cloches and eventually taking them off altogether.

Mini-polytunnels are an alternative to cloches and can be bought from garden suppliers or made from a sheet of plastic and wire arches.

Horticultural fleece is a breathable fabric that is used to protect crops from low temperatures and can be used instead of cloches. It is simply laid over the crops and secured with bricks or stones around the edges. Fleece can be especially useful in protecting crops from flying pests, such as carrot fly, aphids and cabbage white butterflies.

wind

Depending on how exposed your allotment plot is, it may be necessary to create windbreaks to protect your crops from damaging winds, but do check the allotment rules before adding any kind of structure. Solid structures, such as walls or timber fences are not recommended, because they act as a buffer for the wind. The wind will hit the barrier at full force, only to rise above it and then descend in damaging eddies a short distance later.

Hedges provide the best wind protection but take a long time to grow and may not be permitted on the allotment site. Woven willow fences are ideal and come in a range of different heights if your allotment association is fussy about the height of your plot boundaries. Even low, step-over fences will provide protection for low-growing crops. Local wind protection can be provided by cloches or fleece.

birds

Birds are often a nuisance on allotments, particularly the ubiquitous pigeon, which will make a bee-line for cabbages and other brassicas, while many other birds will feast on soft fruit given half a chance. Young vegetable plants are particularly vulnerable to bird damage, so it can help to protect them under a tunnel of chicken-wire or wicker cloches. Later, when plants are bigger and stronger, a light netting can be laid over them. Many allotmenteers find ingenious ways to keep the birds away, including hanging used CDs on strings and making scarecrows out of almost anything that comes to hand. If you are a keen fruit grower and want to devote a large slice of your allotment to soft fruit, it would pay to make or buy a fruit cage.

recycling tip

Cut plastic water bottles in half and use the bottom half as a mini-cloche to protect seedlings from slugs, the weather or animals. The top half can be stuck in the soil, neck end down, and used to get water near the roots of thirsty plants.

supports and structures

Some crops, such as climbing beans, need supports during the growing season. Runner bean frames can be bought from horticultural suppliers, but homemade versions are just as good and much more in keeping with the spirit of the allotment. Bamboo canes are often used to make wigwams or tunnels, secured at the top with wire or string, but these canes are usually imported so are not environmentally friendly. You could always grow your own, but take care because some bamboos are very invasive and may not be permitted on your allotment site. More eco-friendly alternatives to bamboo are coppiced hazel, ash or rowan. Peas also need support, in the form of netting or hazel pea sticks.

Thinking beyond the edible, you could make attractive obelisks from woven hazel, driftwood or anything else to hand (use your creative imagination!) to place in the centre of beds for sweet peas or other climbing ornamentals. Even more ambitiously, perhaps you could construct a wooden pergola for a vine or rose, or a trellis and rose arch to divide your plot in two.

A wigwam support for climbing beans can look decorative as well as maximising the crop.

Wire supports can be used for heavier fruiting plants such as squash or pumpkins to scramble over.

saving seed

Saving seed from your crops is an amazingly rewarding thing to do, but at the same time it is full of intricacies and should be attempted only with a full understanding of how crops are pollinated. It is a complicated subject, and my outline here is very brief, so for those who are interested I would recommend getting hold of a book to guide you further.

The first thing to say is that F1 hybrids should not be used as sources of seed because they have been bred to come true for one generation only, so next time round they may have unwanted characteristics or, indeed, not produce any crop at all. Use older, non-hybrid varieties (known as open-pollinated), whose characteristics you like and that you have found to be reliable, instead. However, bear in mind that plants from the same species can cross with each other, producing mixes of the parent plant, which can destroy the good characteristics in the original

crop and produce plants with unreliable cropping tendencies. Crops can even interbreed with wild relatives – for example, the carrot can cross with the wild carrot (*Daucus carota*), also known as Queen Anne's lace.

On an allotment, with so many other crops growing at close quarters, you could end up with some weird and wonderful mixes unless you isolate the plants to prevent them coming into contact with stray pollen. Because of this, I would suggest that you start off by saving seed only from self-pollinated crops, such as lettuce, peas, tomatoes and French beans (or from flowers, which will present less of a problem if things go wrong). These vegetables self-pollinate before the flower is fully opened, so it is guaranteed that they will not have been cross-pollinated by another source, and will therefore 'come true'.

For more ambitious seed-savers, wind- or insect-pollinated crops can be isolated by bagging the flowers or enclosing the plant in a 'tent' made from an old, fine-mesh net curtain and then hand-pollinating. The easiest of these are the cucurbits – cucumbers, courgettes, squashes and pumpkins – which produce lots of seed from a single pollination, so you need to isolate only one plant. Cover the plant in a mesh tent and hand-pollinate by bringing the male and female flowers together (the female will have the swelling of the nascent fruit behind the flower).

Other plants that produce lots of flowers and require pollination from another plant are trickier to save seed from – for example, carrots, leeks, runner beans, broad beans. On an allotment, I would advise against trying to save seed from these unless you are prepared to risk some strange crosses!

When saving the seed of podded vegetables, leave the pods on the plant for as long as possible to mature and dry.

collecting seeds

Seed of peas and beans are easy to save and collect. Simply leave a few pods on the plant until they are dry and then harvest before the pod splits completely to disperse the seed. For seeds that are produced inside a fruit, such as the cucurbits or tomatoes, leave the fruit to ripen as much as possible. Let tomatoes get over-ripe at the end of the season and then wash the pulp through a sieve so that you are left with just the seeds. Dry on blotting paper. Leave pumpkins, courgettes and squashes on the vine until after the first frosts before splitting them to harvest the seeds. Separate the seeds from the flesh and dry them at room temperature.

storing seeds

Cucurbit and tomato seeds can be stored in paper bags once they are completely dry, but they will need to be kept in a dry place. The fridge is ideal if you can find space. Peas and beans store best in breathable containers, again ideally in the fridge. Although seed can remain viable for several years, for the best results it is better to sow seed you have collected the following year.

A well-organised seed collection, making use of shoe storage bags

storing fruit and vegetables

Once you have started producing your own fruit and vegetables, you won't want to waste any of it. Root vegetables can be stored with some success given the right conditions, and others can be frozen to be savoured in midwinter as a reminder of the long summer months.

root vegetables

Potatoes should be dug up when the ground is dry, and they can then be stored in thick paper sacks, which you should be able to get from an old-fashioned greengrocers. Do not use polythene, which will make them rot. Exclude light from the sack, otherwise they will start turning green.

Other roots, such as carrots, beetroot, turnips, swedes and parsnips, can be left in the ground until the first heavy frosts – and even through the winter if you have a light, free-draining soil. If you lift them make sure they are trimmed and disease free, and store them in wooden boxes filled with damp sand, kept in a cool, dark place. If you have large quantities of roots to store, you could try making a clamp, an old-fashioned method of storing root vegetables either outside or in a dark, cool shed. Pile the vegetables carefully on a thick bed of straw, and then cover them with another layer of straw. Finish off by covering the whole thing with a 15 cm (6 in) layer of earth, leaving a strand or two of straw poking out to allow ventilation.

freezing

The freezer is an invaluable resource when you have a glut, but the types of fruit and vegetable you can freeze are limited. Only freeze those vegetables that you are going to cook afterwards, rather than eating raw. Cauliflower, broccoli, French beans, broad beans and peas can all be frozen successfully, and it helps to blanche them first for 1–2 minutes in boiling salted water. Refresh them afterwards in cold water, and leave them to cool before freezing in airtight plastic bags.

Florets of cauliflower and broccoli can be laid out on trays for fast-freezing, and then transferred to plastic bags. Soft fruit such as raspberries, blackberries and mulberries can also be frozen successfully this way. Other fruit, such as apples, pears and plums can be stewed and pureed before freezing.

Onions should be dried and then hung up in a dry place to be stored. Try plaiting them in the traditional French way.

the directories

This reference section contains all the information
you'll need to see you through the allotment year.
Vegetables are grouped into seven categories –
brassicas, bulb and stem vegetables, fruiting
vegetables, the onion family, legumes, roots and
salads – followed by the fruit and herb directories,
and finishing up with useful illustrated sections on
weeds, pests and diseases.

directory of vegetables

Vegetables are the mainstay of the allotment, and there is a huge and mouth-watering range to choose from. The following directory covers all the most common vegetables suited to a temperate climate, from the humble cabbage to aristocratic asparagus. Quick reference boxes give at-a-glance information, including a maintenance level rating, while more detailed entries on each vegetable will guide you through the growing process from sowing to harvesting.

BRASSICAS

Brassicas, a group that includes cabbage, Brussels sprouts, broccoli, calabrese, cauliflower and kale, are ideally suited to the cool climate of northern Europe, and they are key plants on any allotment. Also included in the family are the brassica root crops, turnips and swedes, but different cultivation methods apply to these, so for the purposes of this book they are grouped with the other root crops (see pages 128–135).

Cultivation

Brassicas are hungry crops and need a rich, fertile and firm soil, tending towards the alkaline (with a pH of 7 or above). They shouldn't be grown in the same place each year, and ideally, in a good crop rotation, they should follow the legumes, which will have left a good, nitrogen-rich soil (see pages 122–127).

Most brassicas are sown into a specially prepared seedbed and transplanted between five and seven weeks later into their

Growing all brassicas under netting or some other protection is advisable, as they are prone to damage from birds and other pests.

final growing positions. This saves space because other, faster maturing crops can be grown before the brassicas are moved. Both the seedbed and permanent bed should be prepared in the autumn by digging in liberal amounts of well-rotted manure or garden compost, as well as adding lime if your soil has a pH below 7 (see page 77). Dig as deeply as possible because brassica roots can penetrate as far as one metre (3 ft).

Sowing

Prepare the seedbed to a fine tilth and water it thoroughly before sowing. Make shallow drills 15 cm (6 in) apart. Seed should be sown as thinly as possible 1.5 cm (½ in) deep.

Transplanting

The seedlings will be ready to transplant five to seven weeks after sowing, when they have three or four leaves and are 10–12 cm (4–5 in) tall. Water the seedlings thoroughly the day before lifting, so that they can be easily pulled out of the soil with minimum damage to the roots. Plant the seedlings as soon as possible into the main bed, using a trowel or hand fork, at the appropriate distance for the brassica concerned. Water them thoroughly after transplanting and continue to water daily until well established. Netting or some other protection may be needed.

Pests and diseases

Clubroot is a fungal disease that manifests itself in the roots of brassicas, distorting and swelling them. Above ground, the leaves may wilt and start turning yellow or brown. If you suspect a plant is suffering from this disease, dig it up and check for signs. If it has clubroot, the whole of the crop will have to be dug up and destroyed. Take special care not to plant brassicas in the same spot for as long as possible, because the spores of clubroot can remain in the soil for up to 20 years. Good crop rotation is the best preventative measure, and because clubroot is more common in acid soils, keeping the soil well limed may help.

Cabbage white caterpillars feast on the leaves of brassicas and can destroy whole plants if left undetected. Yellow, black and hairy, the caterpillars are easy to spot and can be picked off by hand. Grow brassicas under fine netting or horticultural fleece to prevent an attack.

Cabbage root fly lays its eggs in the soil near the stems of brassicas, and it is the small white larvae that do the damage, feeding on the roots. Outward signs are wilting, and damage can be so great that the whole plant dies. They can be deterred by placing a 15 cm (6 in) collar, made from cardboard, old carpet or carpet underlay, around the stems of transplants, and this will prevent the fly from laying eggs on the soil near the plant.

Downy mildew is a fungal infection particularly prevalent in damp springs. Yellow blotches appear on the top of the leaves, with whitish fungal growth underneath. Seedlings are the most susceptible and can be killed. Good ventilation can help prevent the problem, and some cultivars are resistant.

Flea beetle can affect young plants, leaving small holes in the leaves. The tiny beetles jump into the air when disturbed. Protect seedlings and transplants with cloches or horticultural fleece.

Whitefly and aphids

can be sprayed with insecticidal soap. Do this regularly (every few weeks) because these insects have fast lifecycles. Cover the plants with horticultural fleece to prevent attacks from most things.

Birds pigeons especially, are a particular nuisance in winter, and can strip brassica plants bare. Cover plants with fine netting, or criss-cross taught black cotton above the plants to help deter them. Some people spray plants with water that has had garlic boiled in it, but any effect wears off after heavy rain.

Broccoli and calabrese

Brassica oleracea
Italica Group

Broccoli was first cultivated in Italy by the Romans, and it has been grown in Britain since the early 18th century. There are two types: sprouting broccoli produces lots of small florets (available in white or purple); calabrese has one central, much larger floret (usually sold in supermarkets simply as 'broccoli'). Calabrese has a shorter growing season and is ready to harvest in late summer after spring sowing. Sprouting broccoli takes much longer to mature and is ready mid- to late winter the following year.

Soil and situation

Both sprouting broccoli and calabrese require a rich soil that has had plenty of manure or compost added.

Sowing and transplanting

See the general principles for brassica cultivation (page 94–95).

Sow sprouting broccoli under cover in late winter in seed trays or 7.5 cm (3 in) pots, or outside in seedbeds in spring. Sow seed 1.5 cm (½ in) deep in drills 15 cm (6 in) apart, and thin to 10 cm (4 in). Transplant the seedlings to their final positions when they are 10–12 cm (4–5 in) tall, and

SPROUTING BROCCOLI
Sow: indoors in late winter, outdoors in mid-spring
Transplant: when seedlings are 10–12 cm (4–5 in)
Spacing: 60 cm (2 ft) each way
Harvest: late winter the following year
Soil: rich
Maintenance level: high

CALABRESE
Sow: outside in mid-spring
Spacing: 15–20 cm (6–8 in) apart, rows 30 cm (12 in) apart
Harvest: mid- to late summer
Soil: rich
Water: when spears are forming
Maintenance level: moderate

space them 60 cm (2 ft) apart, in rows 60 cm (2 ft) apart.

Unlike sprouting broccoli, calabrese does not take well to transplanting. It is usually sown direct into its final place. Sow seed 1.5 cm (½ in) deep in spring, two or three seeds to a station, 15–20 cm (6–8 in) apart, in rows 30 cm (12 in) apart. Remove weaker seedlings over time, leaving the strongest. Make successional sowings until early summer to prolong the harvest.

Aftercare

Water sprouting broccoli seedlings regularly until well established. Keep the maturing plants free of weeds. To protect from insects or birds, cover with fleece or netting. Plants will be ready to harvest in mid- to late winter the next year. Pick the central spear first, when it is about 10 cm (4 in) long, and continue to pick the sideshoots as they get larger. If the spears are picked every few days, the plant will continue to crop over a period of about two months.

Water calabrese seedlings well, especially in dry summers. Give the plants a good soak a month or so after planting out to improve the yield. Cover with fine netting or horticultural fleece to protect from caterpillars. The first plants will be ready midsummer.

Cultivars

Purple sprouting
- 'Early Purple Sprouting': reliable and tasty, with spears that turn green when cooked.

White sprouting
- 'Early White Sprouting': tall succulent spears, with a cauliflower-like taste.

Calabrese
- 'Romanesco': a decorative, delicious cultivar with light green, pyramidal flowerheads.

Brussels sprouts
Brassica oleracea
Gemmifera Group

Brussels sprouts were first cultivated by the Belgians in the 13th century and subsequently named after their capital city. Winter vegetables, sprouts are traditionally boiled and eaten with Christmas dinner. Sadly they are much maligned, possibly because when overcooked they release sulphur compounds, which can create a bad smell, but when properly cooked they are delicious, with a nutty, flavoursome taste.

Soil and situation
Sprouts do best in a reasonably rich soil, but it shouldn't be over-manured because too much nitrogen can make the sprouts elongate and become loose. Dig in manure or compost in autumn so it has sufficient time to rot down before spring.

Sowing and transplanting
The general principles of brassica cultivation apply (see page 94–95). If you want just a few plants, it's best to sow them in seed trays indoors in late winter. Otherwise, sow them outdoors in a prepared seedbed in mid-spring. Sow 1.5 cm (½ in) deep in rows 15 cm (6 in) apart and thin the seedlings to 7.5 cm (3 in). Transplant the seedlings to their final growing positions when they are 10–12 cm (4–5 in) tall (in early summer), and space them 60 cm (2 ft) apart in rows that are also 60 cm (2 ft) apart. Keep the seedlings well watered until they are well established, and earth up the stems from time to time to make the plants more stable.

Aftercare
Look out for aphids and cabbage whitefly, both of which can be particularly troublesome. If necessary, cover the developing plants with horticultural fleece. The sprouts can be harvested from early autumn to midwinter. Start picking from the bottom of the stem upwards.

Cultivars
- 'Noisette': a French cultivar with smallish sprouts and a delicious nutty flavour.
- Seven Hills': an old cultivar, crops from early to midwinter.
- 'Igor' F1: vigorous 80cm tall plants with tight round sprouts from early to mid winter.

BRUSSELS SPROUTS
Sow: indoors in late winter or outdoors in mid-spring
Transplant: when seedlings are 10–12 cm (4–5 in)
Spacing: 60 cm (2 ft) each way
Harvest: late autumn and over winter
Soil: rich
Maintenance level: moderate

Cabbage
Brassica oleracea
Capitata Group

Cabbage is the most widely grown brassica, related to wild *Brassica oleracea*, a leafy plant native to the Mediterranean, cultivated since Roman times. There are several types of cabbage, which can be harvested

SPRING CABBAGE
Sow: midsummer
Transplant: when seedlings are 10–12 cm (4–5 in)
Spacing: 30 cm (12 in) each way
Harvest: mid-spring
Maintenance level: moderate

SPRING GREENS
Sow: midsummer
Transplant: when seedlings are 10–12 cm (4–5 in)
Spacing: 15 cm (6 in) apart, in rows 30 cm (12 in) apart
Harvest: earlier than spring cabbage, before hearts form
Maintenance level: easy

SUMMER CABBAGE
Sow: early spring (or late winter under glass)
Transplant: when seedlings are 10–12 cm (4–5 in)
Spacing: 35–45 cm (14–18 in) each way
Harvest: midsummer
Maintenance level: moderate

WINTER CABBAGE
Sow: late spring
Transplant: when seedlings are 10–12 cm (4–5 in)
Spacing: 45–60 cm (18–24 in) each way
Harvest: late autumn and winter
Maintenance level: moderate

in different seasons, so if you are fond of cabbage soup, you can ensure a year-round supply. Spring cabbage is sown in midsummer and harvested the following spring; spring greens (also known as collards) are simply spring cabbages planted closer together and harvested earlier, before the hearts have formed. Summer cabbages (both round and pointed cultivars) are non-hardy, fast-growing types: plant in early spring and harvest in summer. Red cabbage is sown in spring, at the same time as summer cabbage, and is ready to harvest in the autumn. Winter cabbages, including the savoys, are hardy cultivars. They take the longest to mature, need plenty of space and good firm soil so they can withstand strong winds.

Soil and situation
Cabbages need good, rich soil, manured in autumn before planting. Lime the soil if the pH is below 7, because cabbages prefer alkaline conditions and do not thrive if the pH is below 7. As they remain for long periods in the soil, they prefer a firm soil.

Sowing and transplanting
The general principles of brassica cultivation apply (see page 94–95). Sow all types of cabbage into a prepared seedbed, or raise seedlings indoors in trays (see page 64). Outside, sow the seed about 1.5 cm (½ in) deep in drills 15 cm (6 in) apart. Thin seedlings to about 7.5 cm (3 in) apart as they mature. Transplant to their final growing positions when they are 10–12 cm (4–5 in) tall.

Aftercare
Water transplanted seedlings generously until well established. If they are troubled by insect pests cover with horticultural fleece or fine netting. Surround them with collars to deter cabbage root fly (see page 95).

Cultivars
Spring
- 'Pixie': used for spring greens or small-hearted cabbage.
- 'Spring Hero': F1 hybrid ball-head type, matures to a good size.

Summer
- 'Derby Day': one of the best ball-head types, light green.
- 'Hispi': a pointed-head cultivar, reliable and with good flavour.

Winter
- 'Christmas Drumhead': well-known, it has a solid heart.
- 'January King': one of the best winter cabbages; the leaves are tinged with red.

Savoy
- 'Vertus': hardy and tasty for midwinter harvest.

Red
- 'Rouge Tête Noir': a good, compact, red ball-head type.

Cauliflower
Brassica oleracea
Botrytis Group

Closely related to broccoli, cauliflowers originated in the eastern Mediterranean and were first recorded in Britain in the late 16th century. Their distinctive white 'curds' (immature flowerheads) are delicious and packed full of vitamin C. They are difficult to grow, needing a certain amount of cosseting and deeply dug, rich soil, as well as sufficient watering, particularly after transplanting and when setting fruit. If the growth is checked, the result is poorly formed, small flowerheads.

There are summer, autumn and winter cauliflowers. Summer and autumn types are easiest to grow; sow in early, mid- or late spring. Winter types are more difficult as they take so long to mature. Sown in late spring, they remain in the ground for almost a year before maturing. As well as standard white cultivars, there are green or purple types. Slightly easier are mini-cauliflowers: grown closer together and harvested earlier.

Soil and situation
Needs a deeply dug, fertile and humus-rich soil for a good crop. Prepare the ground well in autumn with plenty of manure. They prefer a more alkaline soil than other brassicas. Results can be variable with acid soil, even with liberal amounts of lime.

Sowing and transplanting
The general principles of brassica cultivation apply (see page 94–95). Sow summer cultivars into a seedbed in early spring (or sow in autumn for an earlier crop the following year), and autumn types in late spring. Sow seed 1.5 cm (½ in) deep in rows 15 cm (6 in) apart. Thin to 10 cm (4 in). Transplant the seedlings to their final beds when 10–12 cm (4–5 in) tall. Space them at least 60 cm (2 ft) apart, in rows 60 cm (2 ft) apart. Space mini-types 15 cm (6 in) apart each way.

Aftercare
Water seedlings before and after transplanting, keeping as much soil as possible around the roots. Cauliflowers need plenty of moisture to form the heads. Water regularly in the growing season (soak them once a week).

CAULIFLOWER	
Sow: summer types: early spring; autumn: mid- to late spring; winter: late spring	
Transplant: when seedlings are 10–12 cm (4–5 in)	
Spacing: 60 cm (2 ft) each way	
Harvest: late summer to autumn	
Soil: very rich and deeply dug	
Maintenance level: high	

Mulch to conserve moisture. Look out for aphids; spray with insecticidal soap if necessary. Pick off any caterpillars you see.

Cultivars
Summer
- 'Snowball': matures in mid-summer, easy to grow, with tasty, medium-sized heads and large 'wrapper' leaves.

Green
- 'Lateman': widely grown and reliable, it is also suited to growing as a mini-cauliflower.

Autumn
- 'Amphora': an F1, Romanesco type with attractive, lime green, spiralled heads.

Purple
- 'Graffiti': bright purple, a new F1 cultivar with bright purple heads.

Kale
Brassica oleracea
Acephala Group

Kale is a useful low-maintenance vegetable, that crops over winter and into spring, filling the 'hungry gap' when few other vegetables are harvested. It produces bulky, crinkly leaves, which splay out like a palm tree. They are delicious, eaten raw in salads or lightly steamed or boiled, especially when picked young. The young flowerheads, which are borne when the plant starts to bolt in spring, can also be eaten. The Italian black kale, 'Cavolo Nero', is especially decorative and can be used to make the traditional Tuscan soup ribollita.

Soil and situation
Kale is more tolerant of poor soil than some other brassicas, but it will still benefit from a rich soil.

KALE
Sow: outside in late spring
Transplant: when seedlings are 10–12 cm (4–5 in)
Spacing: 60 cm (2 ft) each way
Harvest: late autumn and winter
Soil: will tolerate a poorer soil than other brassicas
Maintenance level: easy

Sowing and transplanting
The general principles of brassica cultivation apply (see page 94). Sow into a prepared seedbed in late spring. Sow seed 1.5 cm (½ in) deep in rows 15 cm (6 in) apart. Transplant to the final growing position when the seedlings are 10–12 cm (4–5 in) tall (in midsummer) and space them 60 cm (2 ft) apart each way.

Aftercare
Water the seedlings in well and keep them free of weeds as they mature. The plants do not need much routine care, but they should be watered from time to time in a drought. They will be ready to harvest in late autumn or winter. Individual leaves can be picked as required.

Cultivars
- 'Dwarf Green Curled': easy and trouble free, producing wide-spreading, curly leaves.
- 'Nero di Toscana' ('Black Tuscany'): interesting and decorative black leaves, which can be shredded into soups.
- 'Russian Red': purplish-grey leaves with reddish ribs; delicious as a cut-and-come-again salad leaf or as kale.

From kale and cauliflower to endless varieties of cabbage, there are many brassicas to experiment with on the allotment.

BULB AND STEM VEGETABLES

From the succulent spears of asparagus to the delicately-flavoured bulbs of Florence fennel, the bulb and stem group includes some of the most prized vegetables of all. On the whole they do best in a rich, moisture-retentive soil so that the stem or bulb can swell unchecked, and, especially in the case of celery and celeriac, they need plenty of water if the weather is dry so that the soil is never allowed to dry out. In the crop rotation, this group falls into the 'Others' list, so they can be grown wherever they can be slotted in.

Growing asparagus requires a good deal of time and care, but it will reward you with crops for many years.

Asparagus
Asparagus officinalis

Nothing beats the taste of fresh asparagus in late spring, especially if you have harvested it yourself, and the ferny foliage can look pretty throughout the summer. But you need a dedicated space for this perennial crop – a bed that can remain undisturbed year after year – and perhaps only if you have a full-size allotment plot should you consider growing it. However, once you have established an asparagus bed, it will go on yielding a crop for up to 20 years, provided you mulch and weed carefully throughout the season.

Asparagus is unusual in that it produces both male and female plants, the female producing fewer, bigger spears but also little red berries in the foliage, which can self-seed. Recently, male-only hybrids with bigger spears have been raised to avoid the self-seeding problem.

Soil and situation
The key to successful asparagus is good preparation of the bed you have earmarked for it. It likes a deep, rich soil, that is also free-draining so dig in plenty of manure and take meticulous care to eradicate all perennial weeds. If your soil is heavy, dig in some extra sand or grit to improve drainage.

Planting
The easiest way to start off an asparagus bed is to buy crowns (root systems) in mid-spring, and for a decent harvest you will need at least 10 plants. Younger crowns (one year old) will transplant more easily, but you will have to wait two years before your first harvest in order to let the plants establish themselves. Older crowns (two or three years old) are also available, and these can sometimes be harvested the following year. Dig a trench that is wide enough for the roots of

the crowns to be spread out, and then make a ridge in the centre. Soak the crowns for a couple of hours before planting, and then place each crown on the ridge, 45 cm (18 in) apart. Cover them with loose soil, to a depth of about 10 cm (4 in). Rows should also be spaced 45 cm (18 in) apart.

Growing from seed

Growing asparagus from seed requires a little more patience, because it will be at least three years before you can start harvesting. Sow into a seedbed in late spring. Sow 1 cm (½ in) deep in drills 30 cm (12 in) apart and thin to 15 cm (6 in). Leave to grow on for a year. Lift the crowns the following year, planting them out as above.

Aftercare

Most importantly, asparagus beds need to be kept weed free, and this must be done carefully and preferably by hand because a hoe might damage the delicate crowns. Mulching regularly with manure can help to suppress weeds and maintain fertility. Watering isn't a priority once the plants are established because asparagus is a Mediterranean crop and fairly drought resistant. The only other thing to be aware of is that asparagus plants do not like being disturbed, so once they are established in their final positions, they are there for good.

Harvest the spears by cutting them in mid-spring, when they are 12–18 cm (30–45 in) high. Cut carefully with a sharp knife about 2.5 cm (1 in) below the soil surface. Traditional curved and serrated asparagus knives are available to buy if you are harvesting a large crop.

Eat your asparagus as soon as possible, steamed and served with butter.

Pests and diseases

Asparagus beetle is, as its name suggests, unique to this plant. The adult beetles are 6–8 mm (¼ in) long and black with yellow blotches on their wings, while the larvae are grey in colour. Both feed on the leaves and bark of the asparagus in spring and summer. Beetles and larvae can be picked off the plants, and affected plants should be burned.

Violet root rot, which can be very damaging, is caused by a fungus that attacks the roots of asparagus. The signs are yellowing, unhealthy-looking fronds in summer. If plants are dug up, you will find purplish-brown filaments spreading across the roots. Remove all affected plants and burn them.

Look out for moles and slugs, which can damage asparagus.

Cultivars

- 'Connovers Colossal': an old cultivar it produces large crops of pale green, flavoursome spears; male and female plants available.
- 'Franklin': an F1, all-male plant with thick, succulent spears.
- 'Jersey Knight': an F1, all-male plant with very high yields; the plump spears are tipped with purple.
- 'Crimson Pacific': a mid-season crop of dark purple spears, tender and delicious.

Celeriac
Apium graveolens var. *rapaceum*

Sometimes known as turnip-rooted celery, celeriac is a close relative of celery, as its name suggests. It is a satisfying winter vegetable, grown for its thickened stem base – delicious grated in a salad or sliced and cooked in soups or stews.

Soil and situation
Celeriac needs a rich, moisture-retentive soil and does best in cool, damp summers.

Sowing
Sow indoors in early spring in modular seed trays because celeriac does not respond well to root disturbance. Grow on in a greenhouse (see page 64). Germination can be slow, and the seeds need light, so scatter them on the surface of the compost rather than covering them.

Transplanting
Harden off the seedlings gradually in a cold frame or sheltered spot outside (see page 67), and plant them out when they are about 7.5 cm (3 in) high in late spring. Set them 30 cm (12 in) apart, in rows also 30 cm (12 in) apart.

Aftercare
Make sure that the ground is kept moist by watering liberally and mulching with straw or well-rotted manure to conserve moisture. As the stem swells, remove the lower leaves to help growth. The plants will be ready to lift in autumn but can be left in the ground until required.

Pests and diseases
Like celery, celeriac can suffer from celery fly, which lays its eggs inside the leaves, causing brown blotches. Remove and destroy badly affected leaves.

CELERIAC
Sow: indoors in mid-spring
Transplant: late spring
Spacing: 30 cm (12 in) each way
Harvest: autumn and winter
Soil: rich and moisture retentive
Maintenance level: moderate

Cultivars
- 'Monarch': a modern cultivar with round, smooth roots, growing to a good size.
- 'Prinz': the large, round roots have an excellent flavour.

Celery
Apium graveolens

Celery can be a tricky and time-consuming crop to grow, and it is not recommended for a drought or dry summer because it needs to be kept permanently damp. Having said that, the modern, self-blanching types are easier than the older trench cultivars, which need to be earthed up, and can produce very good results.

Soil and situation
Celery needs a deeply dug soil rich in organic matter. It must be well drained but also moisture retentive. The plants won't thrive if the soil is too dry in summer.

CELERY
Sow: indoors in early spring
Transplant: late spring
Spacing: 30 cm (12 in) each way
Harvest: late summer to autumn
Soil: rich and damp
Maintenance level: difficult

Sowing

Sow indoors in early spring in a seed tray or modules. The seeds need light to germinate, so scatter them on the surface of the compost rather than covering them. Germination can be erratic, taking up to 2 weeks, so make sure they are at a temperature of 10–15°C (50–59°F), no warmer. Once they have germinated, harden the seedlings off gradually in a cold frame or sheltered spot outside (see page 67).

Transplanting

When they are big enough to handle – when they have five or six leaves – they will be ready to plant. For old-fashioned trench types dig a trench 30 cm (12 in) deep and half-fill it with manure, topping up with soil until you reach a depth of 10 cm (4 in). Set the plants 30 cm (12 in) apart. Self-blanching types are usually planted in a block, with 25–35 cm (10–14 in) each way between each plant.

Aftercare

Celery must be watered thoroughly throughout the growing season, and you must make sure that the soil never dries out. Mulching with straw or well-rotted manure will also help to conserve moisture. Feeding the plants a month or so after transplanting with a liquid feed or high-nitrogen fertilizer can help. Start earthing up trench cultivars when the plants are about 30 cm (12 in) high, and continue to do so throughout the growing season. Harvesting can start at the end of the summer and can last into autumn but it should be completed before the first frosts arrive.

Pests and diseases

Celery is very attractive to slugs, so take appropriate action to avoid them (see page 161). It is also prone to celery fly, whose grubs burrow into the leaves and cause brown blotches in spring. Remove and destroy the damaged leaves.

Cultivars

- 'Golden Self-blanching': an easy, dwarf celery that needs no earthing up.
- 'Tall Utah': a trench celery that needs to be blanched; excellent flavour.
- 'Daybreak': an early maturing self-blanching variety with long, smooth stems. Good bolting tolerance.

Florence fennel
Foeniculum vulgare var. *dulce*

Related to the common fennel (*F. vulgare*) whose leaves and seeds are used for flavouring, this is grown for its fleshy stem base and attractive feathery foliage. The bulb has a pleasant aniseed flavour and is delicious braised or roasted. It needs a hot summer ideally and does best if sown midsummer for an autumn crop. In cold conditions it may bolt, producing flowers at the expense of the bulb.

Soil and situation

Likes dry, sandy soil in open ground but don't let it get too dry.

Sowing

To prevent bolting, sow outside in midsummer after the longest day. Sow seed 1 cm (½ in) deep in drills that are 30 cm (12 in) apart. Thin to about 30 cm (12 in).

FLORENCE FENNEL
Sow: outdoors in midsummer
Spacing: 30 cm (12 in) each way
Harvest: autumn
Soil: light and sandy, not too dry
Maintenance level: moderate

Aftercare

Despite its need for heat, Florence fennel also needs moisture for the bulb to swell, so the plants should be mulched and watered regularly in times of drought. When the bulbs start to develop, they should be earthed up to blanch the stem and keep the plant stable. Harvest in autumn when the bulb is plump, cutting just below soil level.

Pests and diseases

Florence fennel is largely pest and disease free.

Cultivars

- 'Cantino': a modern, bolt-resistant fennel with good flavour.
- 'Zefa Fino': a well-known fennel, ideal for cool climates as it resists bolting; it has large, white bulbs.
- 'Romanesco': classic Italian variety from Rome: it has large white bulbs and good sweet flavour.

Rhubarb
Rheum x cultorum

Strictly a vegetable, but more often regarded as a fruit, rhubarb is an easy-going perennial, that produces delicious, succulent pink stems in spring. Particularly useful for the corner of a plot, once established, it will go on year after year in the same spot, giving useful ground-cover and providing the first 'fruit' crop of the year. For an early crop in winter, rhubarb can be blanched, using forcers, which lengthen the stem a few weeks early.

Soil and situation

Rhubarb is reasonably unfussy about soil and will even grow on an acidic soil. However, before planting, the soil should be enriched by digging in plenty of well-rotted manure or garden compost. Plant in an open situation away from any shady trees.

Planting

Rhubarb is most often grown from sets, dormant rootstocks with at least one bud. Plant in autumn, setting the plants at least 1 m (3 ft) apart. With the buds facing upwards, cover the sets with a thin layer of soil.

Aftercare

Rhubarb doesn't need a huge amount of cosseting, but you will get the best yields by keeping the plants moist in summer and dry in winter. Mulching with grass clippings or well-rotted manure is recommended. Mature plants will send up pretty white flowers in summer, and these can be left, although they may weaken the plant. If you want to force rhubarb for an early crop, cover the crown with straw in late winter, and then cover the whole plant with a terracotta forcing pot or upturned bucket to exclude the light. The rhubarb should be ready to pull about 5 weeks later. For normal harvesting, pull the rhubarb stems regularly, discarding the leaves before cooking. The crowns will need dividing (see page 72) every 5 years or so.

Pests and diseases

Honey fungus, a soil-borne parasite, can affect the roots of rhubarb, manifesting as a whitish fungus around the base of the plant. Weak or old plants are the most susceptible. There is no cure: affected plants must be

RHUBARB
Plant sets: autumn
Spacing: 1 m (3 ft) each way
Harvest: spring and summer
Soil: unfussy, but will do best in a rich soil
Maintenance level: easy

removed, including all traces of root, and burned.

Cultivars

- 'Timperley Early': a well-known cultivar, which is ready to harvest in late winter or early spring; it has long, thick stems and a good flavour.
- 'Victoria': a late maincrop rhubarb, reliable and heavy cropping.

Rhubarb that is 'forced' by covering over the crowns in winter will provide an early harvest of sweet, red stems.

FRUITING VEGETABLES

The fruiting vegetables include everything from tomatoes and chilli peppers to sweetcorn, pumpkins and squash. They are generally quite tender crops, to be planted out later in the season after the last frosts, and tend to catch up with their neighbours quickly as they are all reasonably swift growers. Once again, most fall into the 'Others' group in the crop rotation, so they can either be planted together or mixed in with other crops wherever there is space.

Cucumbers are tender crops so grow them inside and plant out in summer.

Aubergines
Solanum melongena

Even less hardy than peppers and chillies (see page 112–113), aubergines are difficult to grow unless you have a greenhouse, and they must be kept indoors in all but the warmest of weather. Transplant only to a sheltered spot in full sun in summer.

Soil and situation
Grow aubergines in pots or grow-bags in a greenhouse.

Sowing and transplanting
Because aubergines have a long growing period, it helps to get a head start by buying young plants from a nursery or garden centre. However, if you have a heated greenhouse and/or a propagator, you can sow seed in late winter or early spring. Sow 1–2 cm (about ¾ in) deep in a seed tray or sow two seeds to a 9 cm (3½ in) pot, and maintain a temperature of at least 21°C (70°F) until the seeds have germinated. Once they have germinated, gradually reduce the temperature to a minimum of 14°C (57°F). Prick out into individual pots when the seedlings are about 5 cm (2 in) tall, or remove the weaker seedling if you have planted two

seeds to a pot. Pot on as the seedlings fill their pots: a 5 litre (12 in) pot should be the right size for a fully grown plant.

Aftercare
Aubergines need copious amounts of water, particularly if the weather is dry, and they will benefit from a high-potash feed once a week when the flowers are forming. When the plant reaches about 30 cm (12 in)

AUBERGINES
Sow: indoors in mid-spring
Transplant: late spring, into pots or grow-bags
Harvest: late summer to autumn
Soil: rich and moisture retentive
Maintenance level: difficult

high, pinch out the growing tip to encourage bushy growth. Provide support if necessary.

Pests and diseases

The same pests and diseases that affect tomatoes also apply to aubergines (see page 116).

Cultivar

- 'Moneymaker': prolific and reliable, with long, purple fruits.

Courgettes and marrows
Cucurbita pepo

Courgettes are hugely satisfying vegetables to grow because of the speed at which they mature. Almost overnight the fruits begin to swell, and if they are picked regularly the plants will continue to fruit for several weeks. Marrows are really courgettes that have been left to grow

oversized, but for those competitive types growing prize marrows, there are specific cultivars that you can grow from seed. Both are tender vegetables, and should not be planted out until well after the last frosts. Delaying until early summer won't compromise the crop as they are swift growers. Courgettes are available in a wide range of shapes, sizes and colours: the choice is yours.

Soil and situation

Courgettes and marrows need a rich, moisture-retentive soil with plenty of organic matter incorporated into it. An open, sunny site is also required.

Sowing

It is best to sow seed inside in mid- or late spring before planting out in early summer when the nights are warm. Seed can be tricky to germinate, so it can pay to do it in layers of kitchen roll. Dampen a double layer of kitchen paper and individually wrap each seed loosely in a section. Place the seeds in a plastic container, put the lid on and leave in a warm, dark place, such as the airing cupboard. Check after four or five days, and if the seeds have started to sprout, plant them into 7.5 cm (3 in) pots. Alternatively, if temperatures are generally warm, sow each seed directly into a 7.5 cm (3 in) pot, place it on its edge and cover with a 2 cm (¾ in) layer of soil.

COURGETTES AND MARROWS
Sow: indoors in mid-spring
Transplant: late spring or early summer
Spacing: 1 m (3 ft) each way
Harvest: late summer to autumn
Soil: rich and moisture retentive
Maintenance level: easy

The seed will germinate quickly at a temperature of 18°C (64°F) or more – the minimum temperature is 13°C (55°F). Gradually harden off the seedlings (see page 67).

When they have three or four proper leaves, plant them out. Prepare holes for the plants 30 cm (12 in) wide, and fill with well-rotted manure or compost. On top of this pile the backfill, to form a low mound. Plant the seedlings into these mounds. This minimizes stem rot, allowing water to run off the mound. Courgettes and marrows need lots of space, so allow at least 1 m (3 ft) around each plant.

Aftercare

Both crops need plenty of water throughout the growing season and should not be allowed to dry out, however don't overwater in the early days, or leaf production will increase at the expense of the fruit. Give them a good soak when flowering starts and mulch with well-rotted manure to help to retain moisture. Feed once a week, either with a proprietary tomato feed or a comfrey or manure tea (see page 78). Pick courgettes when they are big enough, and keep picking to encourage more fruit. If you're going for a king-sized marrow, pick off all but the best one.

Pests and diseases

Mosaic virus is the main disease associated with courgettes, causing the leaves to appear mottled and yellowy. If left, it can stunt the fruit. Badly affected plants should be destroyed. Botrytis or grey mould can be a problem in wet summers, causing a grey, fluffy mould on leaves, fruit and stem. Remove and destroy all affected parts. Slugs can damage young plants.

Cultivars

Courgette

- 'Defender': an F1 cultivar that is reliable and early, producing masses of regular, glossy green fruits.
- 'Gold Rush': an F1 American cultivar with decorative and tasty bright yellow fruits.
- 'Rondo de Nizza': a round Italian courgette with pale green fruits, which are good for stuffing.

Marrow

- 'Long Green Trailing': dark green marrows with pale stripes; plants need plenty of room to grow.

Cucumber
Cucumis sativus

Cucumbers wouldn't be high on my priority list for an allotment plot unless you have a greenhouse and can devote a substantial amount of time to watering. Both indoor and outdoor types are available, but outside it can be difficult to get a good crop unless you live in a particularly warm and sunny area. Cucumbers do best in warm, humid conditions. Indoor types tend to be smooth skinned, and the outdoor types are warty and rough. There are many cultivars to choose from, including trailing or climbing types. Old cultivars have both male and female flowers, and the male flowers (the ones without a fruit forming behind them) should be removed, because this can result in bitter-tasting cucumbers. Modern cultivars have been bred to have all female flowers, so this needn't be a problem.

Soil and situation

Cucumbers like a very rich soil. Dig in plenty of compost or well-rotted manure before planting out. Outdoors, they need a sheltered, sunny spot.

CUCUMBER
Sow: indoors in late-spring
Transplant: summer
Spacing: 60 cm (2 ft) each way
Harvest: autumn and winter
Soil: very rich and moisture retentive
Maintenance level: difficult

Sowing

Cucumbers are tender crops and shouldn't be planted outside until early summer when there is no danger of frost. For outside types, start seed off indoors in late spring in 7.5 cm (3 in) pots, setting the seed on its side and covering with 2 cm (¾ in) of soil. Greenhouse cultivars can be started earlier. When the young plants are big enough, start hardening them off (see page 67), finally planting them out in early summer. Plant into mounds in the same way as you would do for courgettes and marrows (see page 109), setting them out at least 60 cm (2 ft) apart.

Aftercare

Cucumbers must be watered at least twice a week in summer, otherwise the fruit may not develop properly. Feed with a liquid tomato feed or comfrey tea once a week when the fruits start to swell. Pick the fruits when they are 15–20 cm (6–8 in) long.

Pests and diseases

Cucumbers grown under glass are prone to red spider mite and powdery mildew. Red spider mite affects leaves, flecking them with yellow before they turn brown and die. Keeping the greenhouse humid will help to control the problem. Powdery mildew is a fungal disease that can affect a wide range of crops, coating leaves and stems. Plants that are very dry around the roots are particularly susceptible. All cucumbers can get mosaic virus, which causes the leaves to become mottled and yellowy. There is no cure, so affected plants should be destroyed.

Cultivars

Indoor
- 'Carmen': a modern F1 cucumber with all-female flowers; it is extremely heavy cropping and resistant to powdery mildew.
- 'Telegraph Improved': a traditional, good-cropping cucumber that has been grown since 1873.

Outdoor
- 'Burpless Tasty Green': an F1 hybrid, with smooth-ish fruit with tender skins and good flavour; it can be grown up a trellis.
- 'Crystal Apple': a curiosity, with round yellow fruits on trailing stems.

Globe artichokes and cardoons
Cynara scolymus,
C. cardunculus

Neither of these plants is related to the Jerusalem artichoke (see page 130). The roots of Jerusalem artichokes are eaten, but globe artichokes and cardoons have edible flower buds and leaf ribs, respectively. Globe artichokes are the most widely grown of the three, and are popular in France. If left to mature, the flowers develop into beautiful purple thistle-like heads.

Cardoons are most often grown as ornamental plants, although the fleshy leaf bases can be delicious, especially when blanched. Like Jerusalem artichokes, globe artichokes and cardoons have wonderfully architectural foliage, which can create a dramatic focal point or screen on an allotment.

GLOBE ARTICHOKES AND CARDOONS
Plant offsets: late spring
Spacing: 1.2 m (4 ft) each way
Harvest: late summer
Soil: light and well drained, plenty of organic matter
Maintenance level: easy

Soil and situation

Globe artichokes and cardoons need an open, sunny situation and a light soil that has been enriched with plenty of manure or garden compost. If you have a heavy soil, dig in some sand or sharp grit to improve drainage.

Sowing or planting

It is generally better to grow globe artichokes from rooted offsets as they are extremely variable when grown from seed. These are widely available from mail order seed companies. Alternatively, buy young plants from garden centres or nurseries to get a head start. In late spring plant out the offsets, allowing 1.2 m (4 ft) between them and covering them with a thin layer of soil. Water well until established.

If growing from seed, sow inside in late winter in seed trays, and prick out into individual 7.5 cm (3 in) pots when the seedlings are large enough to handle.

Aftercare

If growing from offsets, the plants may produce small flowers in the first year. Pinch these off to encourage bushy growth. Globe artichokes and cardoons are perennials, so will die back in winter and should be cut back and covered with straw to protect the crowns. They are susceptible to cold, wet soils and are often lost in winter. Because they are short-lived, they should be replaced every three or four years using offsets that the plants produce (see page 72).

To blanch cardoons in autumn, stake the plant and then wrap newspaper around leaves, stem and stake. Tie it all up with twine and after six weeks remove the tender blanched stems. They are tender enough to eat raw, or lightly steamed.

Pests and diseases

Artichokes and cardoons are relatively trouble free, although aphids can be a problem. Use insecticidal soap or spray the stems with strong jets of water to dislodge the aphids.

Cultivars

Globe artichoke

- 'Green Globe': reliable and tasty, it is supposedly Britain's favourite cultivar.

Cardoon

- 'Gigante di Romagna': an old Italian cultivar that is good for blanching and very decorative.

Peppers and chillies

Capsicum annuum Grossum Group, *C. annuum* Longum Group

Peppers and chillies grow best under glass, but can also be grown in pots or grow-bags outside in a sheltered corner, especially if they are against a sunny fence or wall. Like herbs, I count them as 'extras' on an allotment plot – not a staple crop but definitely worthwhile if there is space and you have the time. They are originally from South America, so need a long, warm summer if they are outside.

Soil and situation

Grow in pots or grow-bags, and give them plenty of shelter.

Sowing and transplanting

Peppers and chillies must be sown inside and need a temperature of at least 21°C

(70°F) to germinate. Use a heated propagator if you have one. In early spring sow seed 1–2 cm (about ¾ in) deep in pots or seed trays. Once they have germinated, they need to be kept at a temperature of at least 14°C (57°F). Prick out the seedlings into individual 7.5 cm (3 in) pots when they are 5 cm (2 in) tall, and then over a period of a few weeks gradually lower the temperature to harden them off (see page 67). When the plants are filling the small pots, in late spring, transplant them into grow-bags or large pots. Don't leave them outside until early summer.

Aftercare

Peppers may need some support as the fruits swell, but chillies can be left to their own devices. Both need copious amounts of water, particularly if the weather is very dry, and they will benefit from a high-potash feed once a week once the flowers have started to form. A nettle or comfrey tea will do the job nicely (see page 78). When the plants reach about 30 cm (12 in) high, pinch out the growing tip to encourage good bushy growth lower down.

Pests and diseases
The same pests and diseases that affect tomatoes also apply to peppers and chillies (see page 116).

Cultivars
Peppers
- 'Gypsy': an easy, tasty pepper with tapered fruits that mature from yellow through orange to shiny red.
- 'Purple Beauty': an early variety with fabulous dark, glossy skins, turning red when ripe.

Chillies
- 'Hungarian Hot Wax': one of the easiest chillies to grow, it has long, pointed fruits, which start off yellow and sweet and get hotter as they ripen.

Pumpkins and squashes
Cucurbita maxima,
C. moschata, C. pepo

Pumpkins and squashes are easy and fun crops to grow, and they are particularly suitable for an allotment plot, where they can use the space to spread them-selves around. They come in a range of shapes, sizes and colours, they look attractive and are somehow quite addictive – I would cheerfully give over my entire plot to them. Some are purely decorative, but most are edible, and are delicious roasted or used to make a winter soup. There are two main types: summer squashes, which tend to be smaller, and are treated like courgettes and picked as soon as they ripen; and winter pumpkins and squashes, which mature at the end of the summer and are left outside to ripen and toughen up in the sun where they grew.

PUMPKINS AND SQUASHES
Sow: indoors in mid-spring
Transplant: late spring
Spacing: 1 m (3 ft) each way
Harvest: late summer to autumn
Soil: rich and moisture retentive
Maintenance level: easy

Soil and situation
Pumpkins and squashes have the same soil requirements as courgettes and marrows (see page 109). They need a rich, moisture-retentive soil with plenty of organic matter incorporated, and they should be given an open, sunny site.

Sowing
As with courgettes and marrows, I would recommend sowing seed inside in mid- or late spring before planting out in early summer. Try germinating the seed between layers of kitchen roll before planting in pots (see page 109). Gradually harden off the seedlings (see page 67). When the plants have three or four proper leaves, they can be planted out into mounds, in the same way as courgettes and marrows (see page 109). Allow at least 1.2 m (4 ft) between each plant – or as much space as possible to grow.

Aftercare

Water and feed the plants liberally throughout the season for best results (as with courgettes, see page 109). Summer squashes are harvested in the same way as courgettes – that is, picked regularly over the summer – but pumpkins and winter squashes are ripened on the vine until autumn, their skins left to harden in the sun. If the weather is wet, it can help to lift the pumpkin or squash slightly on a bed of straw. Alternatively, cut the fruit and leave it to finish ripening in the greenhouse. Well-ripened winter squash should keep for many weeks.

Pests and diseases

Both crops are relatively trouble free, although slugs can damage young plants.

Cultivars

Pumpkins

- 'Baby Bear' a small, round pumpkin; which is delicious in pumpkin pies.
- 'Crown Prince': the large, steel-grey fruits have a good flavour.

Summer squash

- 'Custard White': beautiful small, white fruit have scalloped edges.

Winter squash

- 'Butternut': a well known squash with excellent flavour.
- 'Turk's Turban': very decorative with stripy, turban-shaped fruit.

Sweetcorn
Zea mays

Originally from South America, sweetcorn has been grown for thousands of years, and it is a staple crop in many countries. It differs from many other vegetables in that it is wind- rather than insect-pollinated, so it is usually grown in blocks rather than rows so pollination can take place easily. The blocks should consist of at least nine plants, grown in a three-by-three pattern. To make the most of the space on your plot, sweetcorn can be under-planted with other crops, such as squashes or dwarf beans, or with ornamental plants, such as nasturtiums or marigolds.

Soil and situation

Sweetcorn needs a good, rich soil, which should have been deeply dug with organic matter. It is a tender crop, so grow it in a sheltered position in full sun.

Sowing

In milder areas sweetcorn can be sown direct into the ground in late spring when the soil has reached a temperature of at least 12°C (54°F). Warm the soil first if necessary with fleece or cloches. Sow seed 2 cm (¾ in) deep in a block, leaving about 35 cm (14 in) each way. Protect each seed with an upturned jam jar to deter mice. In cooler areas sow one seed to a 7.5 cm (3 in) pot and germinate at a temperature of 15–18°C (59–64°F) in a green-house or light-filled windowsill. Harden the plants off gradually by putting them outside on warm days and bringing them in at night (see page 67). Plant out once all danger of frost has passed, setting them in a block as above. Sweetcorn is susceptible to the cold, so if the weather deteriorates cover the plants with horticultural fleece.

Aftercare

Once they are established the plants will need little watering until the fruits start to swell. At this stage they will benefit from a thorough soaking. On exposed sites the plants should be earthed up to stop them keeling over. Other than this, routine weeding is all that is needed until the end of the summer, when the ripe fruit can be picked. Take care when weeding as roots are relatively near the soil surface.

SWEETCORN
Sow: outdoors in late spring
Spacing: grow in a block, 35 cm (14 in) each way
Harvest: late summer
Soil: rich and deeply dug
Maintenance level: easy

Pests and diseases

Mice can be a nuisance because they are partial to the freshly planted seeds. Place upturned jam jars over the seeds as soon as they have been sown, or bury holly or other such prickly leaves near the seed when planting.

Some people choose to place plastic water bottles over the cobs as they are ripening, particularly if squirrels are a problem in your area. Simply cut out the bottom of the bottle and place over the cob, leaving the bottle top off.

Cultivars

- 'Strawberry Popcorn': a novelty with small, pinkish-red cobs; the kernels can be used for popcorn.
- 'Sundance': recommended for northern climates as it is quick to mature.
- 'Swift': a modern F1 hybrid, which is quick to mature and has exceptionally sweet and tender cobs.

Tomatoes
Lycopersicon esculentum

Everyone likes the taste of home-grown tomatoes, and they are useful for growing in odd corners, either direct in the ground or in pots or grow-bags. They don't take up much space and always look attractive with their shiny red – sometimes orange, yellow or even purple – fruits. Of course, if you have a greenhouse, this will be the natural place to grow them, and you'll probably get a bigger, better crop under glass. Choose either cordon or bush types. Bush tomatoes are easier because they need no support and can be grown in pots or even hanging baskets, whereas cordons need staking and can be trained up trellises or wigwams. Tomatoes come in all sizes, from the smallest cherry tomatoes to the large beefsteak types, and once you have grown your own you will never be able to go back to the tasteless supermarket alternatives, grown for shelf-life rather than taste.

Soil and situation

Grow tomatoes in a sunny, sheltered site in well-drained, fertile soil that has had plenty of organic matter added.

Sowing

Sowing tomatoes from seed is easy, but if you want to cheat and get a head start you can buy small plants from most garden centres. Sow seed indoors in mid-spring. Scatter seed thinly into a 12 cm (5 in) pot and cover the seeds with a 1 cm (½ in) layer of compost and then put a layer of clingfilm on top to conserve moisture. Leave in a warm place, such as an airing cupboard, and watch for them germinating. Once the seedlings have emerged move the pot into the light. When they are large enough to handle prick out into individual 7.5 cm (3 in) pots.

Transplanting

The seedlings must be carefully hardened off (see page 67) before planting out in late spring or early summer, once all danger of frost has passed. If planting direct into the ground, plant cordons 40 cm (16 in) apart, and bush cultivars 50 cm (20 in) apart. Provide support for cordons, either up a single cane, or up a trellis or wigwam. Bush cultivars can be left to their own devices.

TOMATOES
Sow: indoors in mid-spring
Transplant: late spring to early summer
Spacing: 40–50 cm (16–20 in) each way
Harvest: mid- to late summer
Soil: well drained and fertile
Maintenance level: moderate

Aftercare

Bush tomatoes don't need much attention, but cordons are more labour-intensive. The side-shoots on cordons should be nipped out regularly as they grow to encourage upward growth. Use twine to tie in cordons to their stake as they grow. In midsummer, when the cordon has reached its required height, pinch out the growing tip at the top of the plant to stop more flowers forming. All tomato plants must be watered regularly. Those planted in open ground should not need feeding as long as the soil has been enriched with manure or compost, but those in pots or grow-bags will benefit from a high potash liquid feed, such as comfrey or nettle tea, every week or so. The tomatoes can be harvested in mid- to late summer. Leave them on the vine to ripen for the best flavour.

Pests and diseases

Aphids and whitefly can attack tomato leaves, particularly in the greenhouse. Spraying with insecticidal soap can help. Stem rot is another problem caused by spores in the soil, and is most likely to occur when the ground is cold and wet. At ground level, the stem starts to rot, turning brown and sunken, eventually causing the whole plant to wilt. Destroy affected plants.

Blight can also affect tomatoes which, like potatoes, are in the *Solanaceae* family. Most likely to occur in cool, damp summers, the disease is caused by the spores of the fungus *Phytophthora infestans*, and can spread quickly, rotting foliage and destroying whole plants. The first signs are unsightly blotches on foliage and fruit and a mass of creamy white spores on the undersides of the leaves, which spread down into the soil. Plants need adequate space to help prevent the spread of this devastating disease. If affected, remove and destroy all of the affected plants.

Blossom end rot is caused by lack of calcium and is particularly prevalent when there is a lack of water when the fruit first starts to appear. It manifests as an unattractive sunken dark patch on the fruit. There is no cure, but sufficient watering will help to prevent it.

Cultivars

Indoor

- 'Gardener's Delight': a cordon tomato with cherry-sized, sweet and juicy fruit; it can also be grown outside.
- 'Shirley': an F1, medium-sized cordon tomato; reliable and heavy cropping.

Outdoor

- 'Golden Sunrise': a cordon type with bright yellow, medium-sized fruits.
- 'Red Alert': a bush cherry tomato, with high yields and good flavour.
- 'Tigerella': the attractive red fruits are striped with gold when ripe.

Cordon

- 'Black Russian': an unusual beefsteak tomato with purple-red skin, which darkens as the large, slightly irregular fruits ripen.

Fruiting vegetables such as tomatoes, courgettes and pumpkins are some of the most rewarding and decorative crops you can grow on your plot.

ONION FAMILY

The onion family, or alliums, are a relatively small group containing onions, shallots, leeks and garlic. They all require a well-drained, humus-rich soil – not too nutrient-rich – and are easy, low-maintenance crops that gardeners of any level of expertise should have success with. In a four-year crop rotation, the alliums can be given a group of their own (following the root vegetables). In a less complicated three-year rotation, they can be grouped with the legumes, which have similar needs from the soil.

A staple crop on the allotment, onions are extremely easy to cultivate from sets.

Garlic
Allium sativum

All sorts of lore surrounds the humble garlic and its supposed magical properties, and it is also one of the most useful flavouring vegetables in the kitchen. There should be a place for it on every allotment as it is extremely easy to grow. From a single clove planted in autumn, a bulb magically appears, to be harvested the following summer. Garlic actually needs a period of cold in order for the bulb to develop properly, so cool, temperate winters are perfect.

Soil and situation
A light, well-drained soil is ideal, with not much manure added.

Planting
The best time to plant garlic is autumn, before the first frosts, but it can also be planted in late winter. Break the head of garlic up into cloves, and plant the individual cloves so they are upright, pushing them into the soil so that just the tip is showing. Set them out 15 cm (6 in) apart in rows 30 cm (12 in) apart.

> **GARLIC**
> **Plant:** autumn
> **Spacing:** 15 cm (6 in) apart; rows 30 cm (12 in) apart
> **Harvest:** midsummer
> **Soil:** light and well drained
> **Maintenance level:** easy

Aftercare
Garlic can be left to its own devices, although you should weed regularly between plants and rows. Harvest bulbs in midsummer, when the leaves have turned yellow. Don't leave them in the ground because the bulbs might start to sprout again.

Pests and diseases
Garlic can suffer from the same problems as onions (see 120). It can also get leek rust (see 119).

Cultivars
- 'Cristo': large, strongly flavoured bulbs.
- Elephant garlic: really a type of leek, with larger bulbs and a milder flavour.

Leeks
Allium porrum

Leeks, a staple winter allotment crop, have been cultivated for thousands of years. In Roman times they were a favoured vegetable, said to have been eaten daily by Emperor Nero, who claimed that they improved his voice, and they are famous as one of the national emblems of Wales. Normally grown in a seedbed and transplanted when the seedlings reach pencil-thickness, leeks are easy to grow from seed and need little attention throughout the growing season. They also look attractive on an allotment, with their blue-green flags standing out in contrast to the feathery foliage of carrots, for example. Letting a couple go to seed at the end of the season can also look attractive, their round seed-heads resembling the ornamental allium. Choose from early or late cultivars, depending on whether you want an autumn harvest or an overwintering crop that will last you till spring.

Soil and situation
Leeks need a well-dug, well-drained soil that has had plenty of manure or garden compost added. Compacted soil will not give good results.

Sowing and transplanting
Sow seed in mid-spring into a well-prepared seedbed (see page 68). Sow seed as thinly as possible in drills 1.5 cm (½ in) deep and spacing them 15 cm (6 in) apart. Thin if necessary. When the seedlings are about 20 cm (8 in) tall and as thick as a pencil they can be transplanted to their final growing space, perhaps following another early crop that has already been harvested (see page 73). Prepare the bed first by digging over and then, using a dibber, make 15 cm (6 in) deep holes for the seedlings, spacing them 15 cm (6 in) apart in rows that are 30 cm (12 in) apart. Water the seedlings first, and then pull them gently out of the soil with the help of a fork.

LEEKS
Sow: mid-spring
Transplant: early summer
Spacing: 15 cm (6 in) apart; rows 30 cm (12 in) apart
Harvest: autumn to winter
Soil: well drained and humus-rich
Maintenance level: easy

Drop each seedling into a hole and then fill it up with water – there is no need to fill the hole with soil. The seedlings will probably flop immediately afterwards but will soon perk up.

Aftercare
Leeks are unfussy, needing little attention through the growing season. Water sparingly unless there is a severe drought. Harvest from autumn onwards and lift when needed through the winter.

Pests and diseases
The most common problem is leek rust. It manifests as rust-coloured spots and streaks on the leaves. The blanched stem underneath the soil is usually unaffected and can still be eaten. Remove and destroy affected leaves, and do not grow leeks or other alliums in the same spot the next year. Other fungal infections common to alliums can affect leeks (see page 120–121).

Cultivars

- 'Bleu de Solaise': an old French cultivar with blue-tinged leaves; hardy and suitable for early spring harvesting.
- 'Musselburgh': a reliable old-fashioned leek popular for many years.

Onions
Allium cepa

Onions have been cultivated for thousands of years. Valued for their culinary and health-giving properties, they featured in the diets of the Ancient Egyptians and were probably grown even earlier. Today they are ubiquitous, used as the basis of so many savoury dishes that we almost overlook their value. On an allotment they are a staple crop, easy to grow and requiring little maintenance. Try white or red onions, in various sizes, as well as smaller spring or salad types.

They can be grown from seed or from sets (small, immature onions). Sets are the least time-consuming method, but there will be wider choice of cultivar if you buy seed. Sets are more likely to bolt (run up to flower) than seed-sown onions. If possible, choose sets that have been heat treated to kill the flower embryo.

Soil and situation

Onions need a fertile, well-dug soil. Make sure it isn't too compacted, or the bulbs will push themselves out of the soil as they grow. The soil should have had organic matter added to it the previous autumn, but don't grow them on soil that has been recently manured.

Sowing seed or planting sets

Sow onion seed indoors in late winter at a temperature of 10–16°C (50–61°F); overwintering Japanese types should be sown in late summer or autumn. Sow into pots or seed trays and prick out into pots or modular trays when the seedlings are large enough to handle (see page 66). Harden them off gradually in a cold frame or in a sheltered spot outside, until they are ready to plant out in mid-spring. Space them 10 cm (4 in) apart with 25 cm (10 in) between rows. Plant smaller bulbs closer together. Plant sets in mid-spring, pushing them gently into the soil so that the tips are just level with the surface. Space them as for seed-grown transplants. Onions can also be sown in late summer or autumn to over-winter.

Aftercare

Onions don't need much attention during the growing season, but they are sensitive to competition from weeds, so regular hoeing is essential. They won't need much watering, except during extremely dry summers. Onions are harvested at the end of the summer when the foliage turns yellow and starts to fall over. Lift the bulbs and if the weather is dry leave them on the surface of the soil to ripen naturally. If it's wet, cover them with cloches or put them in the greenhouse.

Pests and diseases

Onion fly is the main pest. Small flies lay their eggs on the soil around the base of the onion, and the grubs eat the roots and later the onions themselves. Signs include yellow onion tops and

ONIONS
Sow: indoors in early spring
Plant sets: mid-spring
Spacing: 10 cm (4 in) apart; rows 25 cm (10 in) apart
Harvest: late summer
Soil: fertile and well dug
Maintenance level: easy

wilting plants. There is no effective cure or prevention, although planting parsley between rows is said to deter them. Destroy all affected crops.

A number of fungal diseases can affect onions, particularly if the weather is wet and humid. White rot causes a white, fluffy mould with tiny black spots to grow on the bulbs turning leaves yellow. The black dots are sclerotia, the fruiting bodies of the fungus, which can remain in the soil for up to seven years – take care not to plant any members of the onion family in this part of the allotment for as long as possible.

Downy mildew can affect the leaves, causing pale oval blotches, while onion stem rot turns the bulbs soft, with a grey mould around the neck appearing only several weeks after they have been lifted.

Plants affected with any of these fungal infections should be destroyed as soon as possible to prevent the problem spreading.

Cultivars

- 'Bulldog': uniform golden-brown bulbs; an excellent storer.
- 'Radar': an autumn-planting onion with mild flavour and crunchy texture.
- 'Red Baron': a reliable red onion with a good yield and good for storing.

Shallots
Allium cepa
Aggregatum Group

Shallots are smaller than onions, with a milder, sweeter flavour – and they are even easier to grow. Like onions, they are grown from sets, each set multiplying to produce clusters of six to eight small bulbs, which are lifted in the same way. Maturing faster than onions, they also store for longer.

Soil and situation
Shallots have similar require-ments to onions: a fertile, well-dug soil that has been manured the previous year.

Planting sets
Planting sets is the best option as it gives you a head start. Plant in mid-spring, pushing the sets into the soil so that just the tips are showing. Space them 15 cm (6 in) apart in rows 30 cm (12 in)

SHALLOTS
Sow: indoors in early spring
Plant sets: mid-spring
Spacing: 15 cm (6 in) apart; rows 30 cm (12 in) apart
Harvest: late summer
Soil: fertile and well dug
Maintenance level: easy

apart. Shallots can also be raised from seed in the same way as onions (see page 120).

Aftercare
Like onions, shallots need regular hoeing to keep down the weeds, but they won't need much watering, except during extremely dry summers. Shallots are lifted in midsummer when the tops start to dry out, and they should be harvested in the same way as onions.

Pests and diseases
Shallots suffer from the same pests and diseases as onions (see page 120).

Cultivars
- 'Longor': a traditional French shallot with golden-skinned, elongated bulbs.
- 'Red Sun': a high-yielding cultivar with large, round, red-skinned bulbs; sweet and mild flavour.

PODDED VEGETABLES

The podded vegetables, or legumes, are distinguished from other groups in that they are able to 'fix' nitrogen from the air into the soil via little nodules on their roots. So as well as providing edible beans and peas, at the same time they are improving and enriching the soil by raising nitrogen levels. In a crop rotation, they are followed by the greedy brassicas, which will benefit from the high levels of nitrogen in the soil.

Broad beans are easy to grow and the hardiest of the legumes.

Broad beans
Vicia faba

The broad bean is the oldest of all cultivated beans, having been grown in Europe since the Stone Age. It is the hardiest bean and the earliest to mature, especially if you have sown an early crop in autumn. Once the crop has been cleared, there is still time to use the ground again, preferably for a heavy feeder, such as kale or another brassica that will benefit from the nitrogen-rich soil. Dwarf and taller cultivars are available.

Soil and situation
Broad beans need deep, fertile and well-drained soil to perform their best, although they will grow on poorer soils.

Sowing
Overwintering early beans, such as 'Aquadulce Claudia', can be sown direct into the ground in autumn or into pots that can be left in an unheated cold frame or greenhouse. Outside, cover the emerging seedlings with cloches or horticultural fleece. Sow other cultivars in early spring, as soon as the soil is workable (use cloches if there is still a threat of frost). If sown direct, make sure the soil is well drained (a combination of cold and wet rots the beans). Sow 3 cm (1¼ in) deep in double rows, spacing plants 20 cm (8 in) apart each way and leaving a path 60 cm (2 ft) wide between each double row. You can either take out a drill with a draw hoe or use a trowel or dibber for each individual seed.

Aftercare
After they have germinated, the plants need little attention, although taller cultivars need some support. Erect a post at

BROAD BEANS
Sow: early spring (or autumn for over-wintering cultivars)
Spacing: 20 cm (8 in) each way, in double rows
Harvest: late spring to early summer
Soil: well drained
Maintenance level: easy

either end of the row, with several strings stretched between them. Broad beans can be harvested from early summer, although if you have sown a crop in autumn, it may be even earlier. It's best to harvest broad beans when they are young, otherwise they can be tough and bitter.

Pests and diseases

Blackfly is the main problem affecting broad beans. Technically aphids, blackfly are sap-suckers, and can severely weaken plants and affect yield. They can build up swiftly, doubling their population within a week. They cluster around growing tips, so if the plant is big enough, pinch off the growing tip and take with it its colony of blackfly. Infestations can also be controlled with insecticidal soap. Other problems include chocolate spot, which is a fungal disease that manifests as brownish spots and streaks on the leaves. This is rarer than blackfly and is usually a problem with overwintered beans in a particularly wet spring.

Cultivars

- 'Aquadulce Claudia': the hardiest broad bean and the best for autumn sowing.
- 'Red Epicure': pretty red flowers, green pods and interesting red beans.
- 'The Sutton': a dwarf cultivar, to 30 cm (12 in) high; hardy for early spring sowing.

French beans
Phaseolus vulgaris

Despite their name, French beans originated in South America and were brought to Europe by Spanish conquistadores in the early 1500s. They are a swift-growing and easy crop, and more often than not the only problem is dealing with the glut that occurs in midsummer. They are available in either climbing or bush-forming cultivars, and the pods can range in colour from green to purple to yellow. The most usual way to harvest them is when they are young, when the pods are tender and stringless, but if they are left on the plant to mature, the beans inside the pod can be harvested and dried, when they are known as haricot beans.

Soil and situation

French beans like a rich but fairly light soil, which should have been

FRENCH BEANS
Sow: late spring to early summer
Spacing: dwarf: 7.5 cm (3 in) apart in rows 30 cm (12 in) apart; **Climbing:** 15 cm (6 in) apart
Harvest: mid- to late summer
Soil: light and fertile
Maintenance level: easy

prepared in autumn with well-rotted manure. The soil temperature should be at least 13°C (55°F) before sowing.

Sowing

French beans are quite tender and shouldn't be planted outside until late spring or early summer, well after the last frosts. Sow seed of dwarf cultivars 3 cm (1¼ in) deep with 7.5 cm (3 in) between each plant, in rows 30 cm (12 in) apart. Sow two seeds to a station and later remove the weaker seedling. If you grow climbing beans, erect a support (see page 88) and plant seeds 15 cm (6 in) apart, two to a station. Both climbing and dwarf French bean can also be started off inside in 7.5 cm (3 in) pots.

Aftercare

Earth up the stems of dwarf beans as they grow to provide more support. You could also add twiggy supports as the plants get bigger so they do not topple over with the weight of the beans. The plants should be watered as flowers are starting to form. Lack of water at this time will affect yield. Harvest pods when they are young and succulent, picking regularly to encourage more fruit. For dried beans, leave pods on the stem until autumn. In wet weather, erect cloches over dwarf cultivars. After picking, hang the pods until dry, shell and store the beans in a glass jar.

Pests and diseases
As for runner beans (see page 126). French beans are relatively free from pests or diseases. Halo blight, a bacterial infection, is the only real problem, (see page 168).

Cultivars
Climbing:
- 'Blue Lake': a white-flowering cultivar with green beans; can also be used for haricot beans.
- 'Blauhilde': the beautiful purple pods turn green when cooked.

Dwarf
- 'Opera': a new cultivar, which is a vigorous, reliable cropper.
- 'Purple Queen': waxy, tender purple pods.

Haricot
- 'Barlotta Lingua di Fuoco': an Italian bean with stunning red striped pods, which can be eaten instead if preferred.

Peas
Pisum sativum

Frozen peas are available all year round, but nothing beats the sweet taste of a freshly shelled pea. The so-called garden peas (for shelling) are versatile and easy crops, available as dwarf cultivars, which can be used to form low-growing hedges, or taller cultivars, which climb 1 m (3 ft) or so up their supports. In recent years a series of semi-leafless peas have been bred, with less leaf and more tendril to lessen the need for support. Shelling peas are either wrinkle-seeded or round-seeded. Round-seeded types are hardier and best for earlier sowings. As an alternative, try mangetout or sugar snap peas, of which the pods are eaten too. Peas can be sown successionally, from spring onwards, to give a long cropping season, and both early and late types are available.

PEAS	
Sow:	mid-spring onwards
Spacing:	5–10 cm (2–4 in) apart in wide drills, 50–60 cm (20–24 in) between each drill
Harvest:	summer
Soil:	well drained and fertile
Maintenance level:	moderate

Soil and situation
Peas are easy-going vegetables and will even tolerate some shade. They enjoy cool, damp conditions, but at the same time too much moisture can make them rot, so make sure that the soil is free draining. Add plenty of organic matter to the soil to help it retain moisture.

Sowing
Sow seed outdoors from spring onwards (sowing every three or four weeks if you want successional cropping). Traditionally, peas are sown in wide drills so that the plants grow upwards in a broad row (see page 68). Make a drill 20 cm (8 in) wide and 5 cm (2 in) deep with a hoe or spade, and sow three rows of seed evenly along the bottom of the drill, with 5 cm (2 in) each way. Cover the seeds with soil and tread down lightly.

Aftercare

Because the seeds are attractive to birds and mice, it pays to protect them with a tunnel of chicken wire until the seedlings have emerged and are well-established. Water them well, but take care not to overwater them until the flowers start appearing. Too much water too early can affect the yield. Water thoroughly when the plant flowers, and then again when it is setting fruit. After harvesting, chop down the main plant but leave the roots in the soil. Their nitrogen-fixing nodules will enrich the soil for the next crop.

Support

Most types of pea need some form of support. Traditionally, hazel pea sticks are used, which can be woven in at the top to make a decorative support. Netting supports can also be used, as well as tripods or wigwams. Supports should be pushed into the soil when the seedlings are about 10 cm (4 in) tall and beginning to develop tendrils. The length of support is dictated by the height of the cultivar you are growing.

Pests and diseases

Pea moth is the main pest. The moths lay eggs in the flowers, and the larvae then feed on the peas in the developing pods – often going unnoticed until you shell the peas. The pea moth is a summer pest, so the problem

RUNNER BEANS
Sow: late spring
Spacing: 15 cm (6 in) apart
Harvest: midsummer
Soil: rich
Maintenance level: easy

can be avoided if you are able to make earlier sowings, which flower in early summer.

The other main pests are birds and mice. Traditional methods of deterring mice are to soak the peas in paraffin before planting or to scatter prickly leaves, such as holly (*Ilex* spp.), along the drills before covering the seeds up. To protect crops against birds, cover maturing plants with fine netting or criss-crossed black cotton.

Cultivars
Garden pea
- 'Alderman': a tall pea. Has been in cultivation for many years.
- 'Kelvedon Wonder': an old, wrinkle-seeded cultivar, used as an early or maincrop; quick maturing with dwarf habit.
- 'Markana': a newer, semi-leafless pea, largely self-supporting, grows to 75 cm (30 in).

Mangetout
- 'Sugar Snap': a tall, fast-maturing, heavy-cropping cultivar.

Runner beans
Phaseolus coccineus

Native to South America, runner beans (together with climbing French beans) are one of the most ornamental vegetables you can grow on an allotment. Useful for creating screens, they can also be trained up tripods or over an archway, their cheerful red and white flowers brightening up the plot. Dwarf types are also available. They are fast-growing, easy to cultivate, and high-yielding – even a couple of plants will provide enough beans for a family over the summer. Unlike broad beans, runners are less hardy and won't germinate until the soil is at least 10°C (50°F), so don't plant outside until all danger of frost has passed.

Soil and situation

Prepare a trench in autumn. A deeply cultivated, rich soil will produce the best results.

Dig a trench 60 cm (2 ft) wide and 30 cm (12 in) deep, fork over the bottom and incorporate some well-rotted manure. Over the course of winter the trench can be filled with organic matter – kitchen waste, shredded newspaper, garden compost – topped with soil. If you haven't had a chance to prepare the trench, work in as much well-rotted organic matter as possible before planting.

Sowing

Put up your plant supports (see below) before sowing in late spring. Sow seed 2 cm (¾ in) deep and 15 cm (6 in) apart at the base of canes or sticks. Sowing two seeds to a station lessens the risk of non-germination: the weaker seedling can be removed later. Alternatively, start seeds off inside in 7.5 cm (3 in) pots. Giving seedlings a head start can reduce the risk of slug damage as the plants will be bigger and stronger and therefore more resistant when they are planted outside. Sow dwarf types 15 cm (6 in) apart in rows 60 cm (2 ft) apart.

Aftercare

The first shoots of climbing runner beans may need some encouragement to climb up the support: simply wind them anticlockwise around the pole and tie them in. When they reach the top of the supports, pinch out the growing shoot. Water the plants liberally when they come

into flower. Mulching around the base of the plants will help to conserve moisture and keep weeds at bay. Grow dwarf cultivars as bushes, like dwarf French beans, pinching out the top growing shoot when the plant reaches 30 cm (12 in) high. Harvest the beans from mid-summer, picking regularly to keep the plants fruiting.

Support

There are many ways to support runner beans, depending on how much space you want to devote to them. On a full-sized allotment you will probably have space for a long row, but if you want only a small crop, then a couple of wig-wams will do. See page 88 for ideas on supports.

Pests and diseases

Runner beans are relatively free from pests and disease, although halo blight can be a problem. This is a bacterial infection transmitted through the seed and spread through water droplets. The symptoms are a dark spot in the centre of a paler ring on the leaves. The disease can be controlled early on if you pick off affected leaves, but if it has progressed too far it may kill the whole plant. A more common problem with runner beans is that the flowers are reluctant to set (produce pods). Heavy watering when the flowers begin to appear, and over the course of a week or two can help.

Cultivars

- 'Enorma': a reliable and high-yielding cultivar, with long pods that should be picked often.
- 'Hesta': a well-known dwarf type.
- 'Painted Lady': an old cultivar with well-flavoured pods and attractive red and white flowers.

There are endless varieties of bean to try, from traditional runners (top right) to exotic purple podded varieties (below). Beans that are grown for drying (top left) are known as haricots.

ROOT VEGETABLES

This large group includes the potato, as well as carrots, parsnips, and other similar crops. Roots such as carrots, parsnips, beetroot and radish like a light soil in order for the underground part of the plant to form without being hindered in any way. Potatoes are different in that they need a humus-rich, moisture-retentive soil, as do crops such as turnips and swede, whose round roots need plenty of moisture to swell.

White globe radishes are an alternative to the more usual red varieties. Plant successively over summer months.

Beetroot
Beta vulgaris

Beetroot as we know it today, with its rounded, swollen root, evolved from the wild sea beet found along coasts from India to Britain. In medieval times its root was long, and only the foliage was eaten. It wasn't until at least the 16th century that the round-rooted beetroots were developed. Sweet and velvety, fresh natural beetroot is delicious – with a very different taste from the beetroot we can buy in the shops, pickled heavily in malt vinegar. Easy to grow and fast to mature, beetroot is an ideal catch crop, grown between rows of other vegetables, such as the slower maturing brassicas. There are different colours and shapes as well as the more common red beet, including gold and white cultivars, and even ones with attractive concentric rings when cut. The foliage is also edible: picked young it is great in salads.

Soil and situation
Beetroot enjoys a light but rich soil, so it will grow best in soil that has been well fed over the years with manure. Do not add manure just before sowing.

BEETROOT
Sow: successfully, mid-spring to midsummer
Spacing: 15 cm (6 in) apart; rows 20 cm (8 in) apart
Harvest: summer
Soil: light and fertile
Maintenance level: easy

Sowing
Beetroot can be sown from mid-spring onwards and is best sown successively over the course of the spring and early summer in short rows. The seed will not germinate successfully if the soil temperature is below 8°C (45°F) so if necessary warm up the soil beforehand by using cloches or plastic membrane. Sow 1.5 cm (½ in) deep in rows 20 cm (8 in) apart. The seed is gathered in small clusters, making it quite

easy to handle, so you can place a seed cluster every 2 or 3 cm (1–1¼ in). The seed is programmed to need a good soak of rain before germinating, so soak the seed in water overnight before sowing.

Aftercare

Thin the seedlings as soon as they are large enough to touch each other, leaving about 15 cm (6 in) between each plant. Mulching will help to conserve moisture, but do not overwater except in extreme drought, because this will encourage leaf rather than root growth. Keep the plants well weeded. Harvest early crops when they are about the size of a golf ball. If they get too big they become woody.

Pests and diseases

Generally trouble free, although you may notice leaf spot or discolorations. Beet leaf miner (or the maggots of the leaf mining fly) cause brownish patches on leaves, but won't affect the root.

Cultivars

- 'Boltardy': a bolt-resistant cultivar. Sow early in the year.
- 'Bulls Blood': sweet-tasting and attractive, has deep red leaves.
- 'Burpee's Golden': the gold-coloured roots keep their colour when cooked.
- 'Chioggia': an old Italian cultivar with concentric red and white rings.

Carrot
Daucus carota

The carrot has been cultivated since at least AD 200, when it was mentioned in the writings of Athenaeus. It arrived in Europe in the 14th century from Arabia. Early carrots were either purple or yellow, and it wasn't until the 17th century that orange carrots (containing the pigment carotene) were bred by patriotic Dutch growers wanting to grow the vegetable in the colours of the House of Orange. With its better flavour, the orange carrot soon superseded other types. Today, many cultivars are available, from the short globe types to the long, cylindrical roots of the Nantes carrots. Early cultivars are good for midsummer harvests, or if you like baby carrots. Maincrop carrots take longer to mature and are better for storing but are susceptible to carrot fly.

Soil and situation

The ideal soil is light, deep and free draining, so the roots can swell without restriction. In heavy soil the roots can be misshapen and small, with most of the plant's energy being put into the leaves. A nutrient-rich soil can cause the roots to fork, so do not add organic matter. If you have a heavy soil it is best to choose shorter rooted cultivars.

Sowing

Sow directly into the ground in mid-spring when the temperature of the soil has reached at least 8°C (46°F). Sow seed as thinly as possible 1 cm (½ in) deep in rows 15 cm (6 in) apart. Aim for a seed every 2.5 cm (1 in).

Aftercare

Thinning seedlings should be discouraged, because the smell can attract the dreaded carrot fly. If you do need to thin the seedlings, do it on a damp day, which will minimize the scent.

CARROT
Sow: mid-spring
Spacing: 5–10 cm (2–4 in) apart; rows 15 cm (6 in) apart
Harvest: early to midsummer
Soil: light and free draining
Maintenance level: moderate

Make sure the seedlings are kept weed-free. Once they are established, seedlings need little water. Pull the first carrots in early summer, and harvest as required. On light soils, they can be left in the soil until the first frosts. In heavy soils, they should be pulled and stored (see page 91).

Pests and diseases

Carrot fly is a real problem. The tiny flies are attracted to the roots, where they lay their eggs, and the larvae bury deep into the roots, causing the tops to wilt. Carrot flies are low-flying, and some people believe erecting mesh barriers around the crops can help. Better still, enclose the crop in horticultural fleece. Carrot fly can also be reduced by inter-planting with flowers and other crops, to make it difficult for the flies to find the crop.

Cultivars

- 'Bangor': a maincrop F1 hybrid with uniform, cylindrical roots.
- 'Chantenay Red Cored': a well-known, stump-rooted cultivar, sweet and crisp; good for early or successional sowing
- 'Early Nantes': a Nantes-type carrot with medium-sized, cylindrical roots; fast maturing and excellent flavour.
- 'Flyaway': a modern F1 hybrid that has been bred to be resistant to carrot fly.
- 'Parmex': a globe-type carrot with round, very sweet roots.

Jerusalem artichoke
Helianthus tuberosus

Unrelated to the globe artichoke (see page 111), the Jerusalem artichoke is grown for its tubers, which look like small knobbly potatoes. Sweet and tasty they can be boiled or used very successfully in soups. The plant is tall and leafy, growing up to 3 m (10 ft) high, and can make a good screen or windbreak if planted in a row.

Soil and situation

Jerusalem artichokes are unfussy about soil, but the best, fattest tubers will grow in a rich soil.

Planting

Obtain small tubers from nurseries or mail order firms. Plant in early spring at a depth of 10–15 cm (4–6 in). Leave 30 cm (12 in) between plants if you are planting more than one.

JERUSALEM ARTICHOKE
Plant tubers: early spring
Spacing: 30 cm (12 in) apart in single rows
Harvest: late autumn to winter
Soil: rich
Maintenance level: easy

Aftercare

If you are growing the plant as a windbreak, let it grow without checking throughout the season. Some cultivars produce attractive yellow flowers that look like small sunflowers (to which the plant is related). If you are keen to increase the yield of the tubers, it helps to pinch out the growing shoots when the plant reaches 1.5–2 m (5–6 ft) tall. This will prevent it from flowering and encourage it to put more energy into the tubers. When the foliage and stems die back in autumn, cut it back almost to ground level. Wait until late autumn or winter to harvest the tubers, digging them up as necessary.

Pests and diseases

Slugs can attack the tubers, especially in wet weather.

Cultivars

- 'Fuseau': a newish cultivar with smooth, less knobbly tubers.

Parsnip
Pastinaca sativa

Parsnips have been grown since Roman times. They are extremely hardy, and their flavour is improved by the cold as more of their starch content is turned into sugar. Delicious roasted or made into a soup, parsnips are easy winter crops, and on an allotment you can simply leave them in the ground until you want them.

Soil and situation
Parsnips like a deep, light, yet well-cultivated soil, so the roots are free to develop unchecked.

Sowing
Always use fresh seed because parsnip seed loses its viability rapidly. In mid-spring sow seed 1.5 cm (½ in) deep in drills 30 cm (12 in) apart, sowing two seeds every 15 cm (6 in). If the soil is too cold germination will be slow.

PARSNIP
Sow: mid-spring
Spacing: 15 cm (6 in) apart; rows 30 cm (12 in) apart
Harvest: autumn to winter
Soil: light and well cultivated
Maintenance level: easy

Aftercare
Thin seedlings to a final spacing of 15 cm (6 in) apart or remove the weaker seedling if two seeds were sown to a station. Water well until the young plants are well established. Parsnips don't need much water, but keep the soil sparingly moist, because if you water after the soil has been left to get very dry, they are prone to splitting. They are ready to harvest once the foliage has died down, in mid-autumn, and can be lifted as desired. If there are heavy frosts, they may be difficult to lift. In cold areas either lift before the frosts and store (see page 91) or cover them with straw.

Pests and diseases
Parsnip canker is the main problem, especially in wet weather. It shows as large brown or black blotches on the top of the roots. However, if you choose a modern cultivar that has been bred to resist canker, you shouldn't encounter it. Carrot fly can be a problem (see page 164).

Grow parsnips with carrots and protect with fleece or another prevention method (see page 84).

Cultivars
- 'Avonresister': a modern, canker-resistant cultivar; the short roots are good for shallow soil.
- 'Gladiator': a modern F1 hybrid with good canker resistance and smooth, tapered roots.
- 'Tender and True': a popular, older cultivar, but still with good canker resistance; good-flavoured, long roots.

Potato
Solanum tuberosum

The humble potato is the staple of the vegetable plot. Native to Peru, it has been cultivated for thousands of years but arrived in Europe only around 1570. There are dozens of cultivars available to grow, and they are divided into earlies, second earlies and maincrops, depending on the time they take to mature (earlies 14–16 weeks, second earlies 16–17 weeks and maincrop 18–20 weeks). Potatoes are an easy, low-maintenance crop, and always excellent for an allotment because they provide good groundcover to compete with the weeds. They should always be grown from certified seed potatoes.

Soil and situation
Potatoes aren't fussy crops and will tolerate a range of soils, although they do best in a humus-rich, moisture-retentive soil that errs on the acid, with a pH of 5–6. Dig in plenty of organic matter the autumn before planting.

Chitting
Seed potatoes should be chitted or sprouted in late winter before planting. This is done by placing them in an egg box or seed tray with the 'eyes' uppermost and leaving them in a cool dry place with plenty of light. The potatoes will take about 6 weeks to produce sprouts.

Planting
Plant earlies in spring, and second earlies and maincrops slightly later. For the earliest plantings, make sure the soil temperature is at least 6°C (43°F). The shoots are frost tender, so if the weather gets

colder again, protect them as they emerge with horticultural fleece or cloches. Plant earlies about 10 cm (4 in) deep and 35 cm (14 in) apart in rows 60 cm (2 ft) apart, with the sprouting end uppermost. Second earlies and maincrops should be spaced slightly further apart, 40–45 cm (16–18 in), to allow the plants to spread.

Aftercare
When the foliage reaches about 30 cm (12 in) high, earth up the plants by carefully drawing soil around them with a hoe. This prevents any tubers growing near the surface from turning green. The crop will benefit from a good soaking every two weeks. Tubers will start to form when the plant flowers, and a few early potatoes can be harvested as soon as it starts to flower. Dig the rest as and when you need them, lifting the last few when the haulm (stem and foliage) begins to die back. For late maincrops, you can afford to leave the potatoes in the soil for a couple of weeks after the haulms have died back. This will make the skins tougher and the potatoes will subsequently store for longer.

Pests and diseases
Blight and eelworm are the two most common problems with potatoes. Blight, which is most likely to occur in cool, damp summers, is caused by the spores of the fungus

POTATO
Plant: chitted tubers in early to mid-spring
Spacing: 35–45 cm (14–18 in) apart; rows 60 cm (2 ft) apart
Harvest: early to late summer
Soil: moisture retentive
Maintenance level: easy

Phytophthora infestans. The first signs are brownish patches on the leaves, and then fungal growth on the undersides of the leaves. Infection can also spread to the tubers, turning them brown and sometimes leathery. In the final stages, secondary organisms can hasten the plant's demise, producing a soft, foul-smelling rot. Give plants adequate spacing to help prevent the spread of the disease, and destroy all traces of affected plants. The risk of serious blight can be lessened by earthing up, which prevents the tubers becoming infected, and also by growing only earlies and second earlies, which will be harvested before blight can take hold. There are also plenty of known blight-resistant cultivars.

Potato cyst eelworm can be catastrophic to a potato crop. These microscopic worms lay thousands of eggs in the soil, and they can remain dormant for up to 10 years, waiting for a host plant to arrive, when they hatch and feed on the roots. The first signs are wilting, dying leaves, followed by poor crops. There is no cure – all affected plants must be destroyed.

To avoid both these problems, it is vital that you follow a good rotation plan, so that you do not grow potatoes on the same patch year after year. It is a good idea to consult tradtitional crop rotation plans (see page 49).

Cultivars

First early

- 'Epicure': a frost-resistant potato with good flavour and blight resistance.
- 'Orla': a new cultivar with very good blight resistance.

Second early

- 'Blue Edzell': grow this purplish-blue-skinned potato for its novelty value; it has a floury texture so is best for baking or frying.
- 'Charlotte': a good, waxy salad potato, reliable and well known.

Maincrop

- 'Desiree': a well-known, red-skinned potato with firm, yellowy flesh; good for roasting or baking.
- 'Pink Fir Apple': a late-maturing potato with an interestingly knobbled shape; delicious in salads.

Radish
Raphanus sativus

There are two types of radish, summer and winter, and in Britain small summer radishes are those most commonly grown. Excellent for salads, they are extremely easy to grow and mature quickly, making them a useful catch crop. Larger winter radishes, such as the long white mooli, are popular in China and Japan. These can be boiled or used raw in salads.

Soil and situation
All types of radish grow best in light, sandy soil.

Sowing
Sow summer radishes in early spring and continue making successional sowings in short rows throughout spring and early summer. Sow seed thinly, 1 cm (½ in) deep, in rows 15 cm (6 in) apart. Mooli and winter radishes shouldn't be sown until after the

SUMMER RADISHES
Sow: successionally from early spring
Spacing: 2 cm (¾ in) apart; rows 15 cm (6 in) apart
Harvest: summer
Soil: light
Maintenance level: easy

WINTER RADISHES (MOOLI)
Sow: midsummer
Spacing: 15 cm (6 in) apart; rows 25 cm (10 in) apart
Harvest: autumn/winter
Soil: light
Maintenance level: easy

longest day, in midsummer, otherwise they will run to seed. Sow seed 1 cm (½ in) deep in drills 20–25 cm (8–10 in) apart.

Aftercare
Summer radishes don't need much thinning, but if they are overcrowded thin to 2 cm (¾ in) between each plant. Winter radishes should be thinned to about 15 cm (6 in) apart. Water summer radishes about once a week. Too much water will result in too much topgrowth, while too little water will make the radishes woody. Pull summer radishes as soon as they reach about 2 cm (¾ in) across, which should be within four or five weeks of sowing. If you leave them too long, they will be woody and

tough – although on the other hand, it's not the end of the world if they go to seed as they produce small seedpods that can also be eaten, either stir-fried or in salads. Winter radishes are ready to harvest in autumn, about three months after sowing. They can be left in the ground, although frost and slugs can damage them. It may be better to lift and store them (see page 91), making sure you trim the top-growth so that the root doesn't dry out.

Pests and diseases

Summer radishes are frequently attacked by flea beetles, which make tiny holes in the leaves and weaken seedlings. Covering the seedlings with fleece can prevent attacks. Winter types are susceptible to slug damage as well as clubroot and cabbage root fly (see page 168).

Cultivars

Summer radish

- 'French Breakfast': a well-known cultivar with classic red colour and white tip; a crisp texture and a mild, sweet flavour.
- 'Icicle': small, white cylindrical radishes.

Winter radish

- 'April Cross': an F1 mooli hybrid, with long, white roots.
- 'Black Spanish Round': a traditional round beetroot.

Turnip, swede and kohl rabi
Brassica rapa Rapifera Group, *B. rapa* Napobrassica Group, *B. oleracea* Gongylodes Group

Turnips, swedes and kohl rabi are all members of the brassica family, although it is their root (or in the case of kohl rabi the swollen stem) that is eaten rather than the leaves. Native to Europe, the white turnip has been grown since at least 450 BC, while the swede or rutabaga spread from Sweden into the rest of Europe in the 18th century. Both summer and winter turnips are available, but swede is a winter crop, taking much longer to mature. Kohl rabi is a summer crop, and like the smaller summer turnips, can be grown as a catch crop as it is quick to mature. Like other brassicas all attract many pests (see page 95).

Soil and situation

Turnips, swedes and kohl rabi are cool-climate crops, which need a fertile, damp and cool soil to thrive. Like other members of the brassica family, they do best on alkaline soil, so if necessary apply lime in winter.

Sowing

Sow summer turnips and kohl rabi in mid-spring for a summer harvest. Sow as thinly as possible 1.5 cm (½ in) deep in drills 30 cm (12 in) apart and thin the seedlings when they are still quite small because they grow swiftly. Thin to about 10 cm (4 in) apart or less, depending on whether you

SUMMER TURNIPS AND KOHL RABI
Sow: mid-spring
Spacing: 10 cm (4 in) apart; rows 30 cm (12 in) apart
Harvest: summer
Soil: fertile and moisture retentive
Maintenance level: easy

WINTER TURNIPS AND SWEDE
Sow: late spring
Spacing: 25 cm (10 in) apart; rows 40 cm (16 in) apart
Harvest: autumn to winter
Soil: fertile and moisture retentive
Maintenance level: easy

want baby crops or slightly larger ones. Turnips and kohl rabi don't need much attention once they are established, but growth may be checked if the soil gets very dry, so regular watering can help. Harvest both vegetables when the roots are the size of a golf ball – or at least no larger than a tennis ball.

Sow swede in late spring for harvesting in autumn and winter. Sow 1.5 cm (½ in) deep in rows 40 cm (16 in) apart and thin seedlings to 25 cm (10 in) apart. Maincrop turnips are usually sown in midsummer for harvesting in winter. Sow seed 1.5 cm (½ in) deep in rows 30 cm (12 in) apart, thinning to 15 cm

(6 in). Neither crop needs much looking after, but as with summer turnips, don't let the soil dry out completely, or growth will be checked. Both crops can be left in the ground until they are needed, but all should be lifted by midwinter before the ground is too cold. (See page 91 for storage suggestions.)

Pests and diseases

As brassicas, turnips, swedes and kohl rabi can be affected by the same pests and diseases as the whole family (see page 95).

Cultivars
Summer turnip
- 'Tokyo Cross': fast to mature, with small, round white roots.

Winter turnip
- 'Golden Ball': reliable and tender with yellow roots; good for storing.

Swede
- 'Marian': a well-known modern cultivar with purple skin and golden flesh; it is resistant to clubroot and mildew.

Kohl rabi
- 'Purple Danube': a good F1 hybrid with striking purple flesh and cabbage-like leaves; can be cooked or eaten raw.

Radishes are one of the easiest and quickest root vegetables to grow.

SALAD AND LEAF VEGETABLES

This group is united by a need for plenty of water, which is common sense when you focus on the fact that all the plant's effort is going into producing succulent green leaves. From the myriad varieties of lettuce to the more sophisticated chicory and endive, there are masses of different salads to choose from, with a range of flavours and textures, and many highly decorative as well as tasty. Classified as 'Others' in the crop rotation, they can be grown wherever there is space around your plot.

Leaf beat or chard is easier to grow than spinach and just as tasty.

Chard and leaf beet
Beta vulgaris Cicla Group

Chard and leaf beet are good alternatives to spinach, being easy to grow and generally free from pests and diseases. Used in much the same way as spinach in the kitchen, they are biennials and provided you keep picking, they will produce more leaves throughout the summer, autumn and winter, and even into spring the following year, when they will eventually run to seed. This is in contrast to spinach, which has a tendency to bolt the same season. Leaf beet, or perpetual spinach, is easy to grow, more drought tolerant and very prolific, producing new leaves when picked. Chard is a member of the same family, and is more decorative than leaf beet. Some cultivars have coloured stems and midribs – for example, the bright red 'Rhubarb Chard' or the multicoloured 'Bright Lights' (also known as Rainbow Chard). Stems and leaves can be eaten, either steamed or in stir-fries.

CHARD AND LEAF BEET
Sow: late spring
Spacing: 30 cm (12 in) apart; rows 40 cm (16 in) apart
Harvest: summer to spring the following year
Soil: rich and moisture retentive
Maintenance level: easy

Soil and situation
Like spinach, both leaf beet and chard need a nitrogen-rich, moisture-retentive soil so add plenty of well-rotted manure or compost when digging over the soil in autumn. They will also benefit from soil previously sown with a leguminous green manure.

Sowing

Chard and leaf beet should be sown in late spring. If they are sown too early in the year, they may run to seed the same year. Wait until the weather has warmed up before sowing. Sow 1.5 cm (½ in) deep in rows 40 cm (16 in) apart. Thin to 30 cm (12 in).

Aftercare

Chard is less tolerant than leaf beet and needs to be cosseted in the early stages of growth. Watering is the most important task, but feeding with a liquid feed can help if the soil is poor. Harvest chard and leaf beat regularly, before the leaves get too large, and continue to harvest throughout autumn and winter to keep it cropping until spring.

Pests and diseases

These crops are generally problem free.

Cultivars
Chard
- 'Rhubarb Chard': bright red stems and midribs.
- 'Bright Lights': multicoloured stems in yellow, orange, red and pink.
- 'Swiss Chard': a white-stemmed chard.

Leaf beet
- There are no named cultivars.

Chicory and endive
Chicorium intybus, C. endivia

Chicory and endive are excellent additions to the salad bowl, attractive on the plot and are relatively trouble free. There are three types of chicory: the blanched Belgian or Witloof (white leaf), red chicory or radicchio, and sugarloaf (unblanched white chicory). Endives are related. They are grown like lettuce and have a slightly bitter taste. The curly (or frisee) cultivars, grown in summer, have attractive, frilly leaves, while the Batavians (or escaroles) are broad leaved, upright, and able to withstand colder weather.

Soil and situation

Neither is particularly fussy about soil, but both crops will need a well-drained spot in full sun.

Belgian chicory

Sow Belgian chicory in early to midsummer. Make drills 30 cm (12 in) apart, and sow as thinly as possible 1 cm (½ in) deep. Thin to 20 cm (8 in). In early winter lift the roots and trim leaves to about 2 cm (¾ in) above the neck. Place the roots in a box of sand, and take them out a few at a time. To force, shorten the main roots to about 15 cm (6 in) and plant them upright in a pot in moist soil. Place another pot over the first and store. The resulting chicons take between three and five weeks to develop, depending on temperature.

Radicchio and endives

Sow from mid-spring to midsummer, depending on whether you choose early or late cultivars. Sow 1 cm (½ in) deep in rows 30 cm (12 in) apart, and thin to 25 cm (10 in). Some endives can be blanched, making their hearts more tender and tasty. This can be done simply by laying an upturned saucer over the middle of the plant for a few days before harvesting.

Aftercare

Both crops are easy to please, needing nothing but occasional watering and weeding. Mulching can help to conserve moisture.

Pests and diseases

Chicory and endive are usually trouble free.

Cultivars

Belgian chicory
- 'Witloof de Brussels': the most widely grown cultivar, used for forcing and blanching.

Unblanched chicory
- 'Sugar Loaf': a large, big-hearted chicory, which can be grown under cloches in winter.

Radicchio
- 'Palla Rossa': an attractive salad vegetable with dark red leaves and white veins.

Curly endive
- 'Blond Full Heart': a large cultivar for summer cropping, with green and white leaves.

Lettuce
Lactuca sativa

An invaluable, all-summer-long crop, new cultivars of lettuce offer a wide range of shapes, sizes and colours, and the plants can be extremely decorative as well as tasty. There are four main types: butterheads, which have soft, rounded leaves; crispheads, the largest lettuces with compact hearts; cos or romaine lettuce, with tall, upright leaves; and loose-leaved types, which have divided or crinkly leaves. Choose a selection of different lettuces, and sow successively from spring right through the summer to provide a constant source of fresh leaves for the salad bowl.

Soil and situation
Lettuce grows best in soil that is light and fertile, but also moisture retentive. It is a leafy crop, so needs as much water as it can get. Grow in sun or light shade; too much sun in midsummer can dry the soil out making them bolt.

Sowing
From mid-spring sow direct in the ground, and sow successively until midsummer. Sow seed as thinly as possible in drills 1 cm (½ in) deep in rows 15 cm (6 in) apart. Thin out gradually. Space the small cultivars 15 cm (6 in) apart, and the larger ones 30 cm (12 in) apart. Keep sowing every two weeks for a constant supply.

Aftercare
If the soil is allowed to dry out the plants may bolt, so give lettuce crops plenty of water. The best time to water is in the morning so that the leaves dry out quickly (and are therefore less susceptible to downy mildew). Harvest lettuces as they mature. Pick loose-leaf types

leaf by leaf to use as a cut-and-come-again crop.

Pests and diseases
Slugs and snails love lettuces and can riddle the leaves with small holes overnight. Try protecting the plants with copper rings, with crushed egg shells, or by setting up slug traps (see page 161). Cutworms can destroy young plants. These are actually moth caterpillars that emerge from the soil at night and nibble through the base of young plants. There is no organic alternative but to pick them off at night with a torch – which could be difficult on an allotment! Aphids can also be a problem; use an insecticidal soap if they get out of control. Downy mildew can occur particularly in cool, wet summers. Leaves become yellowy and mottled, and a white downy mould appears on the under-sides. Remove and destroy badly affected plants.

LETTUCE
Sow: successively from mid-spring
Spacing: 15–30 cm (6–12 in) apart; rows 15–30 cm (6–12 in) apart
Harvest: summer
Soil: light and moisture retentive
Maintenance level: moderate

Cultivars
Butterhead
- 'All the Year Round': a well-known and reliable cultivar with a very long season.
- 'Continuity': the distinctive reddish-brown leaves have a good flavour and long season.

Crisphead
- 'Saladin': a reliable iceberg type, with large, well-wrapped heads and a crisp texture.
- 'Webb's Wonderful': a popular cultivar, reliable and long standing.

Cos
- 'Little Gem': an old cultivar that is still extremely popular; small and fast maturing.
- 'Pinokkio': a medium-sized, sweet-tasting cos with textured leaves and yellow heart.

Loose-leaf
- 'Lollo Rossa': a well-known Italian lettuce, grown for its attractive, red-tinged, frilly leaves.
- 'Oakleaf': attractive, deeply lobed, reddish-tinged leaves with a good flavour.

Oriental greens
Brassica rapa varieties

Oriental brassicas are usually grown as cut-and-come-again salad leaves. The plants are sown close together and cut when they are about 10 cm (4 in) high. The group includes pak choi, Chinese cabbage, komatsuna, mizuna, mibuna and mustard greens. Most mail order seed companies will have a section of oriental vegetables, and some offer a dozen or so to choose from, so it's a matter of trial and error to find the ones that suit you and your family.

Soil and situation
Like European cabbages, oriental brassicas need a fertile, nitrogen-rich soil and lots of water.

Sowing
Sow direct into the soil from late spring onwards. Do not sow too early, or the plants might bolt.

For mixed crops, broadcast seed over soil that has been well prepared and raked. Alternatively, sow in rows, scattering seed thinly to a depth of about 1 cm (½ in). Cut-and-come-again crops need little thinning. Thin only if the seedlings are excessively crowded.

Aftercare
Water regularly, which will give higher yields once the leaves have been cut. Use scissors to cut the leaves about 2 cm (¾ in) above the base when they are about 10 cm (4 in) high.

Pests and diseases
Slugs and snails can destroy young seedlings. Protect with collars made from plastic water bottles. Many of the same problems that occur with European brassicas can also affect the oriental types, including the dreaded cabbage root fly (see page 163).

Cultivars
Chinese cabbage
- 'Kasumi': a barrel-shaped cultivar with mid-green leaves and good bolt-resistance.

Pak choi
- 'Joy Choi': dark green leaves and white stalks.

Mustards
- 'Green in Snow': a hardy cultivar, good for winter salads.

Spinach
Spinacea oleracea

True spinach, as opposed to leaf beet or chard, can be quite a demanding crop. It needs regular watering, and as an annual has a tendency to bolt. However, I think its taste and texture are superior to any of the leaf beets, so the extra time and effort are worthwhile. It can be used as a catch crop as it is so quick growing (see page 73). Delicious with pasta or fish, it is usually wilted or steamed, but the young leaves can also be eaten raw in salads.

Soil and situation
Spinach can be grown in light shade, and it prefers cool, moist conditions: a lack of water and too much sun can make it bolt. It is a greedy feeder, needing a nitrogen-rich soil that has been enriched with plenty of well-rotted manure or garden

---

SPINACH
Sow: successively from early spring
Spacing: 15 cm (6 in) apart; rows 30 cm (12 in) apart
Harvest: summer
Soil: rich and moisture retentive
Maintenance level: moderate

---

compost. Being a leafy vegetable, it also needs lots of moisture while it is growing.

Sowing
Sow spinach successively every two or three weeks from early spring onwards. Sow 1.5 cm (½ in) deep in rows 30 cm (12 in) apart. Thin to 15 cm (6 in). Spinach can also be sown in late summer or early autumn for a winter crop; choose cultivars that are particularly winter-hardy.

Aftercare
Regular watering is essential for a good crop of spinach: the soil must never be allowed to dry out completely. Harvest spinach regularly, picking the leaves when they are small for the most tender crop.

Pests and diseases
Downy mildew can affect leaves in cool, wet periods, appearing as a white, furry growth on the undersides of the leaves.

Dispose of affected leaves and thin plants to improve air circulation.

Cultivars
- 'Giant Winter': a hardy cultivar for autumn sowing.
- 'Matador': a summer spinach, which is tasty, vigorous and slow to bolt.

Other salad leaves
Rocket (*Eruca vesicaria*), purslane (*Portulaca oleracea*) and corn salad (*Valerianella locusta*) can all be used as cut-and-come-again salad leaves and treated in the same way as the oriental brassicas. Sow them little and often for a long harvest. Rocket is easy to grow, but has a tendency to run to seed quickly, so cut and sow often. Summer purslane is a half-hardy annual, best harvested when the leaves are young. Corn salad, sometimes known as lamb's lettuce, is hardy and can be used as a winter substitute for lettuce.

Soil and situation
Rocket, purslane and corn salad can all be grown in odd corners to fill gaps on the allotment. They need a rich, moisture-retentive soil.

Sowing
Rocket is best sown successively and in small quantities for a summer-long

harvest. Sow from early spring onwards, either broadcasting seed or sowing 1.5 cm (½ in) deep in drills. Sow purslane in early summer, broadcasting seed for a cut-and-come-again crop, or sowing 1.5 cm (½ in) deep in drills. Corn salad can be sown in late spring (or late summer for a winter crop); it is not suitable for a cut-and-come-again crop because it is slow to re-sprout. Seedlings can be sown indoors and transplanted, spaced 10 cm (4 in) apart in rows 30 cm (12 in) apart.

Aftercare
Watering, weeding and regular cutting of the cut-and-come-again leaves is all that is required.

Pests and diseases
All these salad vegetables are relatively trouble free.

directory of fruit

Fruit usually plays second fiddle to vegetables on an allotment, but growing soft fruit is easy and rewarding, and low-maintenance perennial fruits such as raspberries and currants can be easily slotted into the plan (although you may need a fruit cage or some form of protection from birds). Fruit trees are less often grown but, trained as espaliers and fans apples and pears can be extremely useful for screens and plot dividers performing an ornamental as well as functional role.

SOFT FRUIT

Who could resist a small bed of strawberries, or a cane or two of raspberries on their allotment? Remember, however, that the birds are as partial to the harvest as you are, and if you are serious about growing fruit, it may pay to put up a fruit cage.

Blackberries
Rubus fruticosus

Cultivated blackberries have fruit that is bigger, juicier and more flavoursome than wild brambles, so even if you are surrounded by fields of brambles, it is still worth growing blackberries if you are after a low-maintenance, long-lasting crop. They make useful screens or boundaries on an allotment (particularly the thornless types), and they can also be grown against a north-facing wall or fence, because they will crop in shade. Modern hybrids, such as the boysenberry and tayberry, have been developed by crossing blackberries with other fruit such as raspberries.

Soil and situation

Blackberries prefer a damp soil with plenty of organic matter incorporated into it, and they will grow in sun or shade, provided the ground isn't too dry.

Planting

Plant blackberries in autumn, or any time over winter as long as the ground is workable. If you are growing more than one plant, space them at least 3.5 m (11 ft) apart, because these are large plants that need plenty of space. Cut the plants back to about 20 cm (8 in) after planting to encourage bushy growth. Blackberries need support, so grow them against a wall or fence, or erect post and wire supports, with three parallel wires stretched between two upright posts.

Aftercare

Water in times of drought and mulch each spring with well-rotted manure or compost. Prune in autumn when the plant has finished cropping, cutting out the old canes, as with summer raspberries (see page 149). Tie in the new canes to their supports.

Pests and diseases

Botrytis or grey mould can affect blackberries, particularly when the weather is wet. Pick off and discard the affected fruit. Like raspberries, they can also be affected by cane spot (see page 167).

Cultivars

- 'Fantasia': a relatively new cultivar, discovered growing on an allotment in Surrey; very large, glossy fruit with an excellent flavour.
- 'Oregon Thornless': no spiky thorns and delicious fruit, with the added bonus of decorative, finely cut leaves.

Blackcurrants, redcurrants and whitecurrants
Ribes nigrum, R. rubrum

All three currants are easy to grow and maintain, and when they are grown in bush form they need no support. Blackcurrants are extremely high in vitamin C – just six berries will give you the same amount of vitamin C as a lemon – and have an intense, tart flavour. Blackcurrants are best grown as stool bushes – that is, with branches that rise up directly from the root system. Red- and whitecurrants can be grown as standards or bushes with a leg – that is, a central stem that lifts the bush off the ground, which prevents branches from suckering. All three fruits are high-yielding, so you shouldn't need more than one specimen of each.

Soil and situation

All three do best in an open site, but they will tolerate limited shade. Blackcurrants are heavier feeders than red- or white-currants, so should be planted in soil enriched liberally with well-rotted manure or compost and fed if necessary with a high-potash fertiliser. Too much manure for red- and white-currants can make them grow too fast. All need coolish conditions and will not do well in bone-dry soil.

Planting

Plant bare-rooted stools or container-grown plants in autumn and over the winter if the ground is workable, setting them slightly deeper in the ground than they were in their containers so that the branches divide at soil level. Space plants 1.5–2 m (5–6 ft) apart, depending on size. Dwarf cultivars can be planted closer together. After planting, cut back immediately to 10 cm (4 in) above the ground.

Aftercare

Mulch all currants with well-rotted manure or compost in spring. Blackcurrants are pruned in a different way from red- and whitecurrants, as the latter produce fruit from the old wood; blackcurrants fruit from young, one-year-old shoots. See page 148 for pruning instructions.

Pests and diseases

Blackcurrant gall midge causes leaves to dry and drop off. It can be controlled by using insecticidal soap. Reversion disease causes the buds to swell abnormally, and instead of going on to produce fruit, they shrivel and dry up, sometimes producing a gall. Leaves may appear distorted and smaller than normal. Yield will be affected on a reverted bush. There is no cure, and affected bushes should be dug up and burned. Blackcurrants will also need protection from birds.

Cultivars

Blackcurrant
• 'Ben Lomond': a neat, compact bush, but a heavy cropper; needs minimum pruning.

Redcurrant
• 'Jonkheer van Tets': produces heavy crops of large, flavoursome currants.

Whitecurrant
• 'White Versailles': sweet, yellowish fruit on a strong, vigorous bush.

Gooseberries
Ribes uva-crispa

It's worthwhile making space to grow this easy fruit. Grown as bushes or standards, they are the earliest fruit to crop – you can pick the hard, green fruit in late spring for cooking. Leave some fruit on the bush to ripen naturally.

Soil and situation

Unfussy and easy-going, they don't mind partial shade and need manure or compost only if the soil is light or chalky. They do need moisture-retentive soil – if the soil is too dry, fruits may not form properly.

Planting

Buy two-year-old plants that will fruit in their first year. Plant in late autumn or winter as long as the soil is workable. If you are growing more than one bush, plant them 1.5 m (5 ft) apart. Do not plant them too deeply.

Aftercare

Mulch with well-rotted manure or compost in early spring and water new plants in times of drought. They produce suckers, which should be pulled off in summer when still soft. Like red- and whitecurrants, gooseberries don't need hard pruning; remove a third of the growth after fruiting or in autumn, to open up the centre of the bush to allow air to circulate. See page 148 for detailed pruning instructions.

Pests and diseases

The main problem affecting gooseberries is American gooseberry mildew, a fungal infection usually caused by high humidity. A white, powdery mildew appears on leaves and stems. Cut off and destroy any affected parts. To minimize the risk of getting it, lightly prune the bush each autumn to improve air circulation. Gooseberry sawfly is another potential pest. Easily visible, they are green, spotty, caterpillar-like creatures. Birds can also be a problem, and if you do not have a fruit cage you should net bushes if they attack.

Cultivars

• 'Invicta': a well-known, heavy-cropping gooseberry, with good resistance to mildew and an excellent flavour.
• 'Whinham's Industry': a red cultivar with large, sweet fruit; it is more susceptible to mildew than other cultivars.

Raspberries
Rubus idaeus

Raspberries – particularly the autumn-fruiting types – are useful for a low-maintenance plot, needing little care and attention other than pruning and thinning at the right times of year. They can be planted in a row to form a hedge to divide the plot or as a boundary or screen at one end of the plot.

Soil and situation
Raspberries tolerate light shade and prefer a well-dug, light, moist soil that errs on the acid side of neutral; they do best in soil of a pH 6.5–6.7.

Planting and aftercare
Canes, which are available from garden centres or mail order companies, should be planted in late autumn. Don't plant them too deeply – the bulk of the roots should not be more than about 7.5 cm (3 in) below the surface of the soil – and set them out with 40–60 cm (16–24 in) between each plant. After planting, cut back the canes to about 20 cm (8 in) in order to encourage new bushy growth that will bear fruit. In early spring mulch the canes with animal manure or garden compost and hoe to keep the weeds back. Water if the weather is very dry.

Summer cultivars
Summer-fruiting cultivars produce fruit on year-old shoots. Prune summer cultivars in late summer after fruiting. They will need support in the form of posts and wires or a box system (see page 149).

Autumn cultivars
Autumn raspberries are less demanding, making them a good choice for the low-maintenance plot. They are shorter and stronger plants, so rarely need support, and, because they fruit later, they are less of a target for the birds. Unlike the summer raspberries, which produce fruit on year-old shoots, autumn types produce fruit on the same season's shoots, so prune right down to ground level in late winter, to produce new, fruit-bearing shoots later in the season. Thin new canes gradually through the season as they appear. See page 149 for detailed pruning instructions.

Pests and diseases
Like other soft fruit, raspberries need protecting from birds and squirrels with netting. They can also suffer from chlorosis on soil that is too limey, manifesting as a yellowing of the leaves.

Raspberries are susceptible to three fungal infections. Cane spot causes purplish spots on the stem in late spring or early summer. Cane blight affects older stems at ground level, causing the leaves to shrivel. Spur blight appears on new canes, starting as purplish patches, usually around a leaf bud, which turn black and then grey. These can be treated only with a systemic fungicide. Botrytis, or grey mould, can be a problem in wet summers, manifesting as a spreading, grey mould on leaves and fruit.

Cultivars
Summer
- 'Glen Prosen': a midsummer ripener and prolific cropper; the fruit is firm and delicious.

Autumn
- 'Autumn Bliss': a heavy cropper, it has large berries with good texture and flavour.

Strawberries
Fragaria x ananassa

The intense burst of flavour from a home-grown strawberry is always worth the effort of growing them, especially when compared to the supermarket varieties, which look like strawberries but have half the taste. Finding a corner of the allotment for them shouldn't be too hard. They don't take up much space, although they will need rotating round the plot every three years or so.

Soil and situation
Strawberries need a good, deep, fertile soil with plenty of well-rotted manure incorporated.

Planting
Plant new stock at the beginning of the summer, setting them about 45 cm (18 in) apart in rows 1 m (3 ft) apart. The plants will last about three years before losing their vigour, so before this stage, propagate from them by using rooted runners, which should be planted in late summer in a different bed. Prepare the bed well before planting by incorporating well-rotted manure or compost. Once you have established a new patch, the old plants can be discarded.

Aftercare
Net crops to protect from birds. Spread straw under and around each plant to protect the forming fruits and keep them well aired. On new or young plants, cut off the runners as they form, unless you need to increase stock. After harvesting cut off the old leaves to about 10 cm (4 in) above the crown and remove the old straw. The plants may need watering in times of drought, particularly if you are on sandy or chalky soil, but this should be done only after the plants have flowered.

Pests and diseases
Botrytis or grey mould is a common problem, especially in wet weather, rotting the berries and covering them with a fluffy mould. The only remedy is using a non-organic systemic fungicide. To minimize the problem water the plants in the morning rather than the evening to give them time to dry out.

Powdery mildew can also affect strawberries, manifesting as a dry, whitish powder over leaves, flowers or fruit. Keep the crop well watered, because this often occurs when the roots dry out.

Cultivars
• 'Gariguette': a French cultivar dating from the 1930s; the slightly elongated fruits are juicy and exceptionally sweet.
• 'Pegasus': reliable and prolific, with delicious, glossy fruits.

pruning and training soft fruit

Fruit bushes are an excellent addition for the allotment. The pruning and training process for raspberries, blackcurrants, redcurrants, whitecurrants and gooseberries is relatively simple, and it's worth keeping them in good shape as yield will be much improved. Raspberries and blackberries can be trained along wires as a plot divider or boundary, while currants and gooseberries are grown as individual bushes.

Blackberries

Blackberries should be pruned in the same way as summer-fruiting raspberries, cutting out old canes each year after the plant has fruited. Support the plants on post and wires (see page 149).

As they grow, keep new shoots loosely tied up in the centre of the plant, fanning the older, fruit-bearing canes out on the wires. Once the older canes have been pruned after harvest, new shoots can be untied and re-tied to the wires. Next season, as new shoots grow, repeat the process, loosely bunching them in.

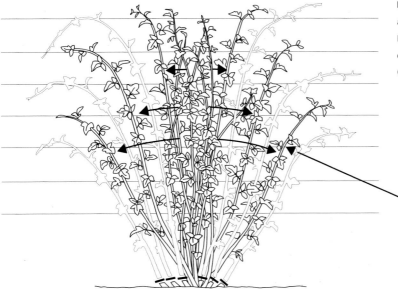

Pruning and training blackberries: Using a post and wire system, keep new canes bunched in the centre, and fan out older canes, loosely tying them to the wires. Cut out old canes at the base in autumn.

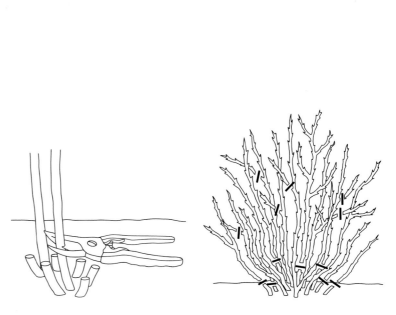

Blackcurrants

These are pruned in a different way from red- and whitecurrants, which produce fruit from the old wood. Blackcurrants fruit from young, one-year-old shoots and should be pruned in late summer, after fruiting. Cut to ground level at least one-third of the old wood – stems that have just borne fruit. The older stems are dark, almost black, while new shoots are paler.

Gooseberries, redcurrants and whitecurrants

These need minimal pruning because they produce fruit from older shoots, but they can still benefit from a light annual pruning to maintain the shape of the plant and maximize air and light circulation. Pruning will also minimize disease and infection and make picking easier. Prune bushes in autumn. Cut back branches by about one-third. Take out old stems until the bush has an open centre.

Pruning blackcurrants (above left): In the first year, immediately after planting, cut back the plant to within two buds or 2.5cm (1 in) above ground level. (Above right): The mature bush should be pruned in late summer, cutting to ground level at least one-third of the old wood, and pruning off any excess side growth to give a good upright shape.

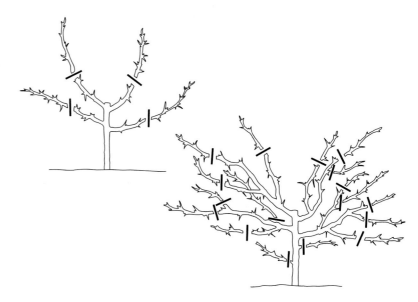

Pruning gooseberries, redcurrants and whitecurrants (above left): With young plants, prune in late summer or autumn, cutting back the main branches to a third of their original length. (Above right): After two or three years, the mature bush should be pruned in autumn, cutting back side-shoots by one-third (leaving about five leaves) to maintain an open centre and vase-like shape.

Raspberries

Summer-fruiting cultivars should be pruned every year in late summer after they have fruited. Cut the old canes as near as possible to the ground and pull new canes out by their roots so that you are left with about five strong stems growing from each original plant. Tie in new canes to their wires, and then, in late winter, cut off the top of each cane just above the top wire so that it does not get too tall. Autumn-fruiting cultivars should be cut back to the ground in late winter, and the new canes thinned gradually through the season as they appear. They should not be pruned at all during the summer months, because fruit is produced on the new shoots.

Autumn raspberries do not need support, because the plants are shorter and sturdier. Summer cultivars can be trained using posts and wires. Use two posts at least 2 m (6 ft) high, and driven about 50 cm (20 in) into the ground, with three wires stretched between, the lowest wire 75 cm (30 in) from the ground. Tie the new shoots into the wires as they grow. Alternatively, use a box system, with two rows of posts 75 cm (30 in) apart, and two wires on each side to enclose the plants. Crisscross twine between them. This obviates the need to tie in new shoots.

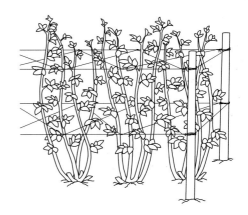

Training summer raspberries (above left): Training on a post and wire system will take up the least amount of space on your plot. The posts should be at least 1.5 m (5 ft) high, with three parallel wires stretched between them, the first about 75 cm above the ground. Tie in the canes to the wires.

This parallel wire system (above right) encloses the canes in a box, lessening the need for them to be tied in. The two rows of posts should be 75 cm (2.5 ft) apart, with two wires along each row. Twine or more wire should be criss-crossed between the parallel wires, to support the canes.

TREE FRUIT

Planting fruit trees may seem like a long-term plan, but there couldn't be anything more delightful on an allotment than a small apple-tree or a series of trees trained as cordons or espaliers. Here we focus on apples and pears, which are more likely to be found on an allotment plot, but you could try plums, damsons or greengages too.

Apples
Malus domestica

Think small when you are buying fruit trees for an allotment, because many sites will not permit trees over a certain size. If you decide to make the long-term commitment to an apple tree, the most important thing when buying is to get the right rootstock, which will determine the eventual size of the tree (see page 152). Most apples are cross-pollinated, which means that to produce fruit they need to be fertilized by pollen from another, similar, tree. This means that you should plant at least two trees that flower at the same time; a good nursery will advise on suitable cultivars. Apples are hardy trees and flower in late spring, usually when all danger of frost has passed, so they are easy to grow in the temperate British climate.

Soil and situation
Apples need a sheltered, well-drained site in ground that never gets too dry. If your soil is light and sandy dig in plenty of well-rotted manure or compost.

Planting fruit trees
Plant bare-rooted specimens in autumn, to give them time to settle in before the growing season. Dig a large hole and mix in plenty of well-rotted manure or compost and a handful or two of bonemeal. Place a stake in the hole, and then plant the tree, making sure that the join of the graft is above soil level. If it is below, the graft could send out its own roots. Fill in first with compost and then with soil, firming it down as you go with your feet. Water in thoroughly.

Aftercare
If the spring weather is dry, water well while the tree is establishing, and add a generous layer of mulch around the base of the tree once the soil has warmed up. A single specimen tree needs little pruning,

but trained trees need pruning carefully in summer (see page 152). If an excess of fruit develops, thinning the fruit as they form can be beneficial.

Pests and diseases

Apples are prone to canker, which can kill a whole tree. Symptoms are distorted, swollen or sunken patches on the bark, sometimes with red or white raised spots, as well as shrivelled leaves. Apple scab is a common fungal infection of both apples and pears. Brown, scabby patches appear on the leaves and later the fruit. Finally, the tiny white and brown caterpillars of codling moths can play havoc with the fruit. They are often not discovered until the fruit is picked. The moths lay their eggs on the leaves and fruit in midsummer, and the caterpillars burrow into the fruit, often through the stalk area so infestation is difficult to detect. Try pheromone traps to control the moths (see page 164).

Cultivars

- 'Cox's Orange Pippin': a well-known apple with deliciously sweet, small fruit.
- 'James Grieve': a sweet and prolific eating apple.
- 'Egremont Russet': a good choice for growing as a cordon.

Pears
Pyrus communis

Pear trees flower earlier than apples, so they are at risk from frost damage, and in colder areas they can be disappointing. Like apples, they can be trained as cordons, fans or espaliers, and a mixed screen of apple and pear cordons looks delightful. Pear trees are usually grafted on to quince rootstock, to produce a manageable tree; Quince C is recommended for cordons or espaliers. Most pears are cross-pollinated, so unless you choose a self-fertile cultivar, you will need to plant two trees.

Soil and situation

Pears need a sheltered, warm spot, with protection from the cold and plenty of sun to ripen the fruit. They aren't too fussy about soil, although they won't do well on excessively alkaline soil.

Planting

As for apples (see page 150).

Aftercare

As for apples (see page 150).

Pests and diseases

Birds can be a real pest in spring, damaging the flowerbuds. Pears are less prone to disease than apples, but they can suffer from canker, pear midge and pear leaf blister. Pear midges lay their eggs in the unopened blossoms, and the resulting larvae feast on the newly formed fruitlets, which will develop cracked patches, eventually turning black, and usually falling to the ground. From here, the larvae go into the soil to pupate, starting the cycle again. If you notice damaged fruit, remove and destroy it, and place a barrier, such as plastic sheeting, on the soil around the tree to prevent more infestation. Pear leaf blister is caused by mites that feed on the leaves, causing yellow, red or brown blisters, seen on both sides of the leaf. Fruiting is not affected.

Cultivars

- 'Beth': an early pear, maturing in late summer; good flavour and reliable cropping.
- 'Beurre Hardy': a mid-season pear, one of the finest dessert cultivars with juicy, sweet fruit.
- 'Conference': the most popular cultivar in Britain, ripening in mid-autumn.

pruning and training apples and pears

Large fruit trees are not often permitted on allotment plots, but you may be able to train them into various shapes to create attractive boundaries for your plot. Apples and pears lend themselves to this sort of treatment, and don't necessarily need a wall to be trained against. There are numerous shapes to prune apples and pears into, but here I focus on just three – cordons, espaliers and step-overs – because of their suitability for allotments. None of them needs a fence or wall to train the plant against, but all provide some sort of boundary or screen – and an attractive one at that.

For all three, it is best to buy one-year-old, single-stem whips or slightly older feathered maidens (young trees that have produced many sideshoots). Alternatively, buy partly trained plants from specialist nurseries. Make sure you buy spur-bearing types (those that produce shoots and fruit buds from short woody clusters called spurs) rather than tip-bearing cultivars. This shouldn't be difficult as most apples and pears are spur-bearing.

Rootstocks

Fruit trees are usually grafted on to rootstocks, which will govern the rate at which they grow. Dwarfing rootstocks are usually used for trees to be trained into cordons, espaliers or step-overs, but if you have poor soil, they may not succeed. Good nurseries will offer a choice of rootstock and advise which to buy, depending on what you are planning. For apples, M9 is the recommended rootstock for standards or half-standards, growing 2–2.4 m (6–8 ft) high. M27 is even smaller, but this can be difficult to grow in all but the best soil. M26 is recommended for trained trees, such as cordons or espaliers.

Cordons

A cordon is a single stem trained to grow at an angle of 45 degrees with numerous fruiting spurs, because this restricts the growth of the leader (central stem) and increases the fruit-bearing potential. A series of several cordons planted along parallel wires makes a good boundary or windbreak. Plant the whip into a hole prepared with well-rotted compost or manure, in front of a series of three taut horizontal wires attached to sturdy uprights. For the first two or three years,

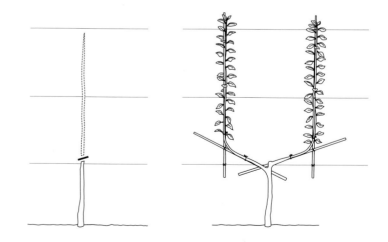

For a double cordon, cut a young, single-stemmed tree to 25 cm (10 in), just above two good buds (above left). As the two shoots grow, tie them into short canes, the first two at a slight angle, and the second two vertical (above right).

before the tree produces fruit, tie in the leader at a 45 degree angle and prune the sideshoots (or laterals) to three leaves beyond the basal leaf cluster – to about 7.5 cm (3 in) – throughout the summer. Any new shoots that appear from this pruning should be trimmed to one leaf. Once the leader has reached the desired height, prune it back to two buds at the end of summer each year. Double and multiple cordons can be created, using the same pruning and training principles but with more stems, grown upwards and parallel to each other (see diagram opposite).

Espaliers

Espaliers can be trained onto horizontal wires attached to fences or on to free-standing wires attached to posts, as with cordons. Either start from scratch with a young, untrained whip or buy partly trained espaliers from a specialist nursery. Plant the tree into a well-prepared hole and tie the laterals to the wires. Espaliers may have two, three or four parallel branches, depending on the height you wish to achieve. You can create more side branches by cutting out the leader about 45 cm (18 in) above the last tier. New shoots will develop, and you can leave one to grow up while those on either side can be tied to the wires. Throughout the growing season tie down new growth on the side branches. In late

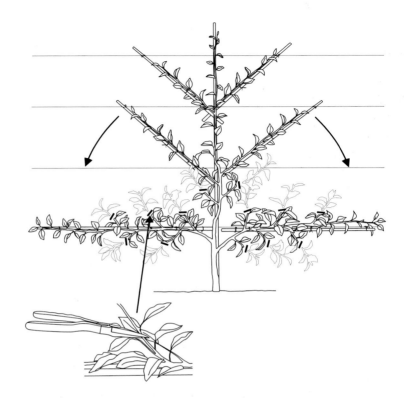

To train an espalier, tie in the laterals to the wire on either side of the leader. To create another tier, cut out the leader 45cm above the last tier, and train two new shoots along another wire, leaving a third to grow upwards as the leader. For quick growth the laterals can be tied in to canes at an angle at first, and then trained horizontally when long enough. In summer, prune out any surplus shoots on the laterals to three leaves from their base. Prune any shoots from the spurs on the leader to one leaf.

summer prune back any shoots appearing on the branches to three leaves from their base, removing any surplus shoots growing from the leader.

Step-overs

Step-overs are really mini-espaliers, trained with only one set of side branches growing about 30 cm (12 in) above the ground. Pruning and training principles are the same as for the espalier.

directory of herbs and flowers

Herbs are a valuable addition on an allotment plot. They fill odd corners and are decorative as edging for beds or paths. It is also fun to grow edible flowers – nasturtiums or marigolds, for example or borage with its star-like flowers that are added to Pimms.

Chives
Allium schoenoprasum

Chives make an ornamental edging to a bed, with their pink-purple pompon flowerheads in summer. The mild, onion-flavoured leaves are used to flavour salads and potato dishes. They prefer fertile, moist soil and full sun. Sow seed in a pot in spring and plant out 10 cm (4 in) apart in late spring; or divide existing clumps of bulbs and replant.

Dill
Anethum graveolens

The yellow flower-heads and feathery leaves are attractive on any allotment plot. A perennial, it is grown for its leaves often used to flavour fish, and its seeds, with their pleasant aniseed flavour. Broadcast seed in a sunny, open spot, mid- to late spring. Harvest leaves when plants are strong. To harvest seeds cut flowers when turning brown and shake, upside down, over a sheet of paper.

Chervil
Anthriscus cerefolium

Similar to dill, chervil has a mild aniseed flavour and soft, ferny foliage. It grows fast and can be harvested only six weeks after sowing. Good for filling a spot in light shade, its seed should be broadcast in mid-spring and again if necessary over the summer.

French tarragon
Artemisia dracunculus

Often used for flavouring chicken, this is a better culinary choice than Russian tarragon with a superior flavour. Like mint, it spreads by underground rhizomes and can be quite rampant. It likes a sheltered, sunny site and well-drained soil. It can't be propagated from seed, so either get hold of a piece of root or rhizome or buy a plant.

Borage
Borago officinalis

Some gardeners regard borage as a weed as it will self-seed freely in conditions it likes. But despite its rather coarse foliage and rangy stems I, for one welcome its starry blue flowers, used in salads or summer drinks. The leaves have a cucumbery taste, and are added to Pimm's. Broadcast seed in spring on open ground. Thin to 30 cm (12 in) apart.

Coriander
Coriandrum sativum

Used widely in Thai cooking, the leaves have a delicate and delicious taste, while seeds can be crushed and used as a spice for curries. Sow seed in an open, sunny spot in early summer and don't overwater. Plants have a tendency to bolt so it can be better to sow seed at intervals over the summer to provide a good quantity of leaves. To harvest the seeds cut flowers when they brown and place upside down, over a sheet of paper to catch the seeds.

Marigold
Calendula officinalis

These are a favourite on many allotments because they can play so many roles. The sunny orange flowers look cheerful and pretty edging beds, and attract many beneficial insects. The petals can be eaten, and used to colour and flavour rice or salads. They are easy to grow – simply broadcast seed in early spring – and in an open sunny spot they will self-seed prolifically to ensure their return the following year.

Fennel
Foeniculum vulgare

Fennel grows readily in well-drained soil and full sun, and self-seeds everywhere given half a chance, so watch out for unwanted seedlings. Tall and willowy, it has fresh, ferny foliage and golden flowers in mid-summer. The foliage is used as a flavouring, especially with fish, and the seeds, like dill, have an aniseedy taste. It is traditionally used as a digestive. Sow seeds in the ground in early summer.

Lovage
Levisticum officinale

Lovage can grow up to 2 m (6 ft) tall. It likes a rich, moist soil in sun or light shade. The leaves taste a bit like celery and can be used to flavour soups and casseroles. Sow seed indoors in early spring and plant out early summer, or buy plants from a nursery or garden centre. The plants take three years to reach full size, and you really only need one plant because of their size.

Mint
Mentha spp.

Beware when growing mint because it can take over, so be mindful of where you plant it. Grow a few species in their own bed, or contain a plant by keeping it in its pot. Propagate from root cuttings or buy small plants from a nursery or garden centre. There are several species and numerous cultivars to suit every need.

Marjoram
Origanum spp.

There are many members of this genus, but sweet marjoram (*O. marjorana*) has the best taste

and a good, bushy habit. The leaves are used in Mediterranean cooking to flavour sauces and soups and can be used either fresh or dried. Sow seed indoors in early spring and plant out in early summer in an open, sunny position. Alternatively, buy plants rom a nursery or garden centre.

Basil
Ocimum basilicum

A tender herb, sow outdoors in early summer. Choose from an array of types from lemon basil, to Thai and Greek basil, all are delicious. Start seed off indoors on a window-sill in mid-spring. Sow thinly into pots or trays and thin out into individual 7.5 cm (3 in) pots. Plant out early summer when temperatures are high, spacing them 30 cm (12 in) apart. Basil is the main ingredient in pesto and is invaluable when preparing Italian dishes.

Parsley
Petroselinum crispum

Parsley is widely grown in Britain and is a staple herb for flavouring and garnish. Curly and flat leaf types are available, both needing a good, rich soil to thrive. Sow thinly from early spring through the summer and water well. Cut off flowers as they appear.

Sorrel
Rumex spp.

Popular in French cuisine, sorrel has a sharp, almost lemony taste. It can be used in soups, purées and salads. It is an easy plant for well-drained ground in sun or light shade and can be grown from seed. Sow indoors early spring and plant out after all danger of frost has past. It has a tendency to run to flower, so pinch out the flowering stems when they appear to encourage more leaf growth.

Rosemary
Rosmarinus officinalis

Aromatic rosemary forms large bushes, and is clipped to make domes. Plant as a boundary for a herb garden or separate area. As a culinary herb it is often used to accompany lamb. A shrubby perennial, it is difficult to grow from seed, so buy plants from a nursery or garden centre. Plant in an open, sunny well-drained site.

Sage
Salvia officinalis

Sage is another useful culinary herb, traditionally used in stuffings to accompany pork. Forming mounds of soft, felty leaves, with blue flower spikes in summer, it is an attractive plant, especially the many cultivars, such as 'Purpurascens', with purple-grey leaves, or 'Tricolor', which has pink- and white-edged leaves. Sages must have light, free-draining soil and plenty of sun; they do not thrive in heavy soils. They can be propagated easily from cuttings.

Nasturtium
Tropaeolum majus

Nasturtiums scramble freely over open ground or they edge a bed, with their pretty orange, red and yellow flowers brightening up a plot. Attracting beneficial insects, the flowers and leaves can be added to salads. Plant seed 1 cm (½ in) deep in spring, in an open, sunny spot.

Thyme
Thymus spp.

There are many different types of thyme with different habits, from creeping thymes to more upright species and cultivars, any of which could be grown on an allotment. The tiny leaves have a delicate flavour, used widely in Mediterranean cuisine, and because they originate from that region, they need Mediterranean conditions: well-drained, dryish soil and full sun. Thymes can be propagated by taking softwood cuttings in spring and summer. Plant out the following spring, incorporating some grit into the planting hole to aid drainage.

problems, pests and diseases

Everyone with an allotment has their share of problems in the form of insects, birds, animals or diseases, but with good practice you should be able to minimize them. Sticking to a good crop rotation will reduce the chances of soil-borne diseases, and keeping the soil healthy with plenty of organic matter will make your plants stronger and more resistant to attacks. Weeds can sometimes harbour disease, so keeping on top of these will also help. Finally keep your plot as diverse as possible. By planting flowers among your crops, you will attract the creatures that prey upon the baddies.

WEEDS

The following perennial weeds are the main offenders on allotments around Britain, and I suggest ways of tackling them without resorting to chemicals. Annual weeds must be controlled too, of course, but they are much easier to deal with, either by using a hoe to cut off the tops or by pulling them out by hand.

Perennial weeds

Bindweed
Convolvulus arvensis

One of the commonest of perennial weeds, bindweed is a strong, twining climber. It can have a huge root and rhizome system, rapidly spreading through the soil and extending to a depth of several metres. An initial blitz – clearing topgrowth and digging out as much root as possible – will allow you to start planting. Keeping on top of it is the only answer. Hoe off new growth and dig up every little piece of root you can find. With persistence, it is possible to eradicate bindweed in about

two years, although new colonies very often arrive from neighbouring plots or land.

Ground elder
Aegopodium podagraria

This common weed is fast spreading and invasive, multiplying by its underground stems. Luckily, the rhizomes do not penetrate far into the soil and can be relatively easily forked up. The main problem is that it regenerates very easily from the tiniest pieces of rhizome, so it is extremely difficult to eradicate. Smothering the plant with black plastic will weaken the plant, but this could take three or more seasons to be effective.

Couch grass
Elymus repens

This weed can look like any ordinary field grass, but the roots are incredibly invasive, forming large networks of underground stems. If the plot is completely covered with couch grass, it is best to wait until late autumn or early winter, when the root system is at its weakest, and then systematically cut squares of turf with a sharp-edged spade or mattock. Make sure that sections of root below the surface are also removed. The turf can be piled up and left under a layer of cardboard or carpet until it has rotted down.

Horsetail
Equisetum arvense

Horsetail is one of the most dreaded and persistent weeds, with roots that are usually too deep to dig out. Repeated chopping back or mowing will weaken the rhizomes, but this takes time. Deep digging is the best solution, although the roots penetrate right into the subsoil, making life very difficult. Cover the area with black plastic or carpet to exclude the light and eventually weaken the plant.

Bramble
Rubus fruticosus

As everyone knows, brambles have their armour above ground. The long, arching, prickly stems can extend 3 m (10 ft) or more, with four or five stemming from one root. Chopping this back can be hard work, and ordinary shears and scythes are often not tough enough. Some people recommend heavy-duty tools, such as machetes, mattocks or azadas (a Spanish agricultural tool that looks like a cross between an axe, a hoe and a machete!) Cutting back should be followed by digging up the roots.

Dock
Rumex spp.

Another common perennial on allotments, dock has a long taproot, so make sure you dig up every last bit of the root, or it will re-sprout. Dig up when they are young and the task will be easier.

Perennial nettle
Urtica dioica

Easily recognizable from the yellowish tangles of roots that run on underground rhizomes, nettles are actually a sign that the ground is fertile. The topgrowth can also be used to make a nutritious plant tonic by soaking them in a bucket of water.

Annual weeds

Annual meadow grass
Poa annua

Meadow grass is common on allotments where there are grass paths running through the site. As an annual it is easy to pull out. It can also be controlled by heavy mulching. It seeds freely, so make sure you catch it before this stage.

Hairy bittercress
Cardamine hirsuta

Related to watercress, hairy bittercress can be eaten when young. As with other annuals, it is easy to pull up or hoe off.

Fat hen
Chenopodium album

The succulent leaves can be eaten, and the greenish-blue flowers are not unattractive. Easily controlled by pulling or hoeing.

Groundsel
Senecio vulgaris

The familiar yellow flowers produce hundreds of seeds, so make sure you pull or hoe it before it sets seed.

Common chickweed
Stellaria media

This is a particularly hardy and persistent weed, germinating even in autumn and flowering in winter in mild places. It can spread quickly if it is allowed to set seed, but it can be controlled by hand-weeding or hoeing.

Speedwell
Veronica persica

Although speedwell has pretty, jewel-blue flowers, it can spread quickly and become a nuisance on an allotment.

ANIMALS AND BIRDS

From slugs and snails to thieving birds, there are plenty of pests to contend with on an allotment. Many plots are adjacent to open countryside, opening up opportunities for larger animals such as foxes, badgers and even deer. But on a positive note, all sites are havens for wildlife, and we should learn to live with the animals rather than waging war on them – perhaps with the exception of slugs and snails!

Rabbits

Rabbits are probably the main pest, and they can inflict serious damage, particularly to young plants. If rabbits become a problem and the site as a whole is open to them, you may have to enclose your plot entirely within a low wire fence. The wire mesh should be buried to a depth of at least 20 cm (8 in) into the soil to stop them burrowing underneath.

Mice

Mice can be irritating, and they are partial to peas and beans. A traditional method of deterring mice is to place holly leaves along the trenches as you plant seeds. Other animal visitors may include foxes, badgers, moles and deer, but the main pests are far smaller and more persistent than both of these and invariably of the insect or bird family.

Birds

Birds, especially pigeons, can be a real nuisance on allotments, devouring soft fruit, such as strawberries, and pecking away at brassicas, particularly the young plants with succulent leaves. Protecting plants can be done in a number of ways. It will pay to build or buy a fruit frame if you are growing lots of soft fruit. Young plants can be protected in a number of ways, from wicker cloches to chicken wire tunnels. Buy horticultural fleece or netting or stretch black cotton between sticks above the plants, to deter birds.

Slugs and snails

Slugs and snails are a perpetual problem on allotments, and especially in damper areas or in wet weather. What many people don't realize is the main damage comes from small, black-keeled slugs, which nibble away at many crops. Every seasoned allotmenteer will have tried different methods of eliminating these creatures, from beer traps to crushed eggshells round the base of plants. Some swear by copper rings, which let out a small electric charge that deters slugs. You could also try using nematodes, a biological control, which are watered into warm soil. Encourage natural predators, such as hedgehogs, by making dry, woody habitats for them.

INSECTS: GOOD AND BAD

Not all insects are bad news in the vegetable garden. To sort the good from the bad, this list will help you identify which should be encouraged, and which should be destroyed.

Friends

Ground beetles

Beetles will eat slugs and caterpillars, but they are also partial to fruit, so keep your eyes peeled if growing strawberries.

Centipedes

Centipedes are partial to slugs, which is a definite boon in the vegetable garden. They will also eat a number of other pests and will only nibble at plant roots in times of drought. You can encourage them on to your plot by creating areas for them to shelter – an old pile of logs, for example.

Lacewings

Common lacewings have large, delicate wings. The larvae are brown and bristly with six legs and a long tail. Adults and larvae feast on aphids and also eat small caterpillars. They are attracted to areas of plant diversity, so planting lots of flowers among your vegetables will help. You can also buy lacewing 'hotels' – special chambers in which they will overwinter.

Hoverflies

Hoverflies are an excellent control for aphids. They lay their eggs in aphid colonies so the grubs have a ready meal waiting for them. Plant marigolds among your vegetables to attract them.

Ladybirds

Common ladybirds get throuh a huge number of aphids. Adults are easily recognizable, but even before they pupate the spiny black and orange larvae consume huge numbers of pests. You can buy ladybird boxes, where they can spend the winter.

Spiders

Spiders consume almost any insect, including wasps and springtails. Provide a good home for them by laying down straw mulch or planting shrubs.

Ichneumon wasp

A large, slender wasp with four wings, it lays eggs inside caterpillars. When the eggs hatch, the caterpillar dies, providing a tasty meal for the young wasps.

Foes

Cabbage white caterpillars

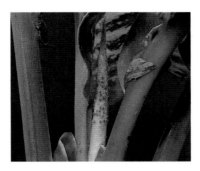

Yellow, black and hairy, the caterpillars feast on the leaves of brassicas and can destroy young plants if left undetected. Pick them off by hand.

Celery fly

Celery fly lays its eggs on the soil around celery and related plants, and tiny white grubs burrow into the leaves, causing blotching and browning. The plants are most vulnerable when young. Remove and destroy any affected leaves.

Aphids

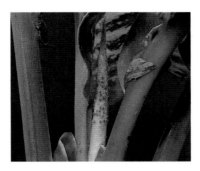

Small but infuriating, aphids are one of the most common pests in the garden in spring and summer. There are more than 500 varieties in Europe alone, in colours, including black, red, yellow and green. The female-only population breeds prodigiously – aphids can reproduce after a week – so they spread like wildfire. Greenfly is a universal pest affecting a number of plants. The insects themselves cannot inflict much damage, although they can make a plant's growth weaker. The main problem is that they are sapsuckers, meaning they spread viral diseases. Blackfly, which often infest broad beans, can

be controlled by pinching out the tips of plants. Natural predators, such as ladybirds, lacewings and hoverflies, can help control aphids (see page 162). Insecticidal soap can help control both.

Asparagus beetle

These black and yellow beetles are easy to spot and remove, but if they go undetected, they can completely strip a plant of its foliage. They are active from late spring to late summer.

Cabbage root fly

Affecting all brassicas, particularly recent transplants, this is a devastating pest, that can completely destroy young plants. The grubs of the cabbage root fly feed on the roots of brassicas, weakening them and causing complete collapse if the plant is young. Deter the flies from laying their eggs on the soil around the plant by placing collars, made of cardboard or old carpet, around each plant or by covering them with horticultural fleece.

Carrot fly

These tiny flies are easy to miss, but it is the grubs that do most damage, eating away at the roots and causing the tops to wilt. The peak danger time is early summer. Adult flies cannot fly above about 75 cm (30 in), so in theory crops can be protected by a fence of mesh at least this high. It is probably safest to cover the crop in horticultural fleece.

Codling moths

These moths lay their eggs on the leaves and fruit of apple trees. The maggots that hatch burrow into the fruit, usually near the stalk, so it is difficult to spot them until it is too late. They feed on the core area, and the apple is rotten when picked. Catching the moths before they lay their eggs is the only way to control them, and this is done with pheromone traps. These contain pellets that mimic the scent of the female moth and lure males into the traps. This won't completely clear the problem, but it will help.

Gooseberry sawfly

Small, green caterpillars with black spots eat the leaf tissue of gooseberry bushes, reducing the leaves to skeletons. The only organic method of control is to be vigilant and pick them off by hand.

Flea beetles

Flea beetles attack brassicas of all kinds, particularly in dry periods in late spring and early summer. They make tiny holes in the young leaves. Seedlings and young transplants can be very badly affected. Protect young plants with horticultural fleece and make sure that the ground does not dry out.

Leatherjackets

These large, grey-brown grubs are the larvae of the crane fly (daddy-long-legs) and feed on brassicas roots as well as other vegetables. They can be controlled by parasitic nematodes, which are watered into warm, moist soil.

Cutworms

Cutworms are the larvae of moths that lay their eggs around the stems of various plants in summer. The resulting greenish, curled caterpillars remain just under the surface of the soil and eat through the stems of plants, sometimes severing them completely. To discourage the egg-laying, surround plants with circles of felt or carpet underlay.

Eelworm

Microscopic worms lay thousands of eggs in the soil, and these can remain dormant for up to 10 years, waiting for the appropriate host plant, often potatoes, to arrive. When it does, they hatch out and feed on the roots. There is no cure – all affected plants must be destroyed and you will have to avoid that patch of ground for potatoes for as long as possible.

Onion fly

Like small houseflies, onion flies lay eggs on the soil around onion or leek crops. These hatch into small white maggots, which feed on the roots and then on the rotting bulbs. The first symptoms are wilting, yellowy leaves. If these pests attack young plants, whole crops can be lost. There is no cure or treatment, so affected plants must be removed and destroyed.

Pea Moth

This is a common pest of peas that is only really a problem between June and August.

The moths lay eggs in the flowers, and the larvae then feed on the peas in the developing pods - often going unnoticed until you shell the peas. As the pea moth is a summer pest, the problem can be avoided if you are able to make earlier sowings, which flower and fruit in early summer.

Pear midge

The midges lay their eggs in the unopened blossoms of pear trees, and the resulting larvae feast on the newly formed fruitlets. The fruit will turn black and usually fall to the ground. If you notice damaged fruits, remove and destroy them. The larvae pupate in the soil beneath the tree, so prevent further infestations by placing a barrier, such as plastic sheeting or cardboard, on the soil around the tree.

Red spider mite

Mainly a problem in green-houses, outdoor strawberries and cucumbers can be affected in midsummer. The mites are tiny and can be seen only if you use a magnifying glass. They produce threads like a spider's web, affecting leaves and flecking them yellow and brown, eventually killing the whole plant. Keeping the greenhouse humid can help,

as can introducing a biological control, *Phytoseiulus persimilis*, which will feed on the mites.

Pear leaf blister

This is caused by mites that feed on the leaves, causing yellow, red or brown blisters, which can be seen on both sides of the leaf. Fruiting is not affected, but you should remove affected leaves as soon as you see them.

Whitefly

Whitefly is a problem inside the greenhouse but also for brassicas throughout the summer period. They are found on the underside of leaves and can severely weaken a plant. The flies lay the eggs, which turn into small, sap-sucking grubs that finally hatch into flies. In the greenhouse they can be controlled by sticky traps, or by the parasitic wasp *Encarsia formosa*. Outside, on cabbages, use insecticidal soap.

DISEASES

American gooseberry mildew

A common fungal disease, this affects gooseberries and blackcurrants. In late spring patches of white fungus appear on new shoots, spreading to the leaves and then the fruits. The patches on the leaves eventually turn brown. If the infection is severe, fruit production will be affected. Prune out and burn affected branches and prune carefully to improve air circulation through the plant. Some cultivars have been bred to be resistant to mildew, and if you are not growing organically use a fungicidal spray.

Apple and pear scab

This is the most widely spread fungal disease of apples and pears. The symptoms are brown, scabby blotches, which appear first on the leaves and later on the fruit. Infected leaves and fruit may fall prematurely. Although unattractive, the fruit is still edible. Collect and burn the leaves and fruit that have fallen and prune out infected branches. If you are planting new trees, choose cultivars that show some resistance.

Beet leaf miner

The maggots of the mangold or leaf mining fly burrow into the leaf surfaces of beetroot or leaf beet, causing brownish patches on the leaves. On beetroot, the edible root is not affected. Pick off leaves that are infected and tread on the larvae.

Blight

Blight affects both tomatoes and potatoes. Caused by the spores of the fungus Phytophthora infestans, it is most likely to occur in cool, damp summers, when the spores are spread by wind and rain. For potatoes, the first signs are brownish patches on the leaves, and then fungal growth on the undersides of the leaves. The stems may turn brown, and the whole of the aerial part of the plant can collapse. Infection can spread the tubers, turning them brown and sometimes leathery. In the final stages, secondary organisms can hasten the plant's demise, producing a soft, foul-smelling rot. With tomatoes, brownish lesions appear on the leaves and stem, and the fruit turns brown and leathery. Give plants plenty of space to help prevent the spread of the disease. Destroy all traces of affected plants, especially potatoes. Infected tubers often lurk unseen in the ground.

Blossom end rot

This is not strictly a disease, but a disorder caused by lack of calcium (resulting from inadequate watering), and it can affect tomatoes and relations such as aubergines and peppers. The symptoms are light brown spots, which appear at the flower end of the fruit and eventually form a dark, leathery patch, making the fruit look flattened. It can be prevented by giving the plant sufficient water when the fruits are forming. If the soil is too dry, nutrients won't be absorbed properly.

Botrytis
grey mould

This common disease affects a wide range of plants – ornamentals as well as vegetables. Strawberries and raspberries are particularly susceptible, as are tomatoes, cucumbers and courgettes, especially if they are grown in a greenhouse. Caused by the fungus *Botrytis cinerea*, it manifests as a swiftly spreading, grey, fuzzy mould, which can attack all parts of a plant. It can spread quickly in humid conditions, as the mould produces airborne spores, which can infect healthy plants through wounds or lesions, or soft, ripe fruit. There is no cure. Make sure that you clear up all infected matter and give plants enough space to improve air circulation and limit the spread of the disease.

Cane blight

This fungal infection of raspberries starts at ground level, but the first symptom you are likely to notice is shrivelling leaves on the older canes. If you see these, check for dark, cracked patches on the canes near the ground. Infected canes should be cut out below ground level. The fungus enters the canes through wounds, so take care when you are weeding around canes.

Cane spot

This is a common fungal disease affecting raspberries and its relatives. It starts as small, purple spots on the canes, appearing in late spring. The size of the spots will increase, so that they eventually look like shallow, whitish marks with a purple edge. The canes may die back, and fruit production will be affected. There is no cure for affected canes, so they must be cut out.

Canker

This serious disease affecting apples and pears, causes distorted, swollen or sunken patches on the bark, and sometimes red or white raised spots, or shrivelled leaves. Fruit can also start to rot. The spores of the fungus enter through scars or lesions on the bark, leaf or bud of the trees. Prune out affected branches and peel away affected bark until you are left with pale, healthy tissue. Choose canker-resistant cultivars.

Chocolate spot

Chocolate spot is caused by the fungus *Botrytis fabae*, and it affects broad beans, particularly in damp weather. Round, dark chocolate-coloured spots appear on the leaves and stems, and in severe cases these spots can merge and blacken, resulting in the death of the plant. However, in most cases, it will not be fatal, and the crops can still be harvested. It may result in fewer pods. Burn infected plants after harvest.

Clubroot

This is a serious soil-borne infection common to brassicas, and it can be devastating. The spores of the fungus *Plasmodiophora brassicae* attack the roots of cabbages and related vegetables, sometimes resulting in the collapse of the aerial part of the plant. The roots become distorted and swollen, while the plants wilt and appear stunted. All affected plants must be dug up and destroyed. The bad news is that the fungal spores can live in the soil for up to 20 years, so you should not plant any brassicas in the area as long as possible. Liming the soil can help, because clubroot is more likely to occur in acid soils.

Downy mildew

As with all fungal infections, downy mildew is most likely to be seen in wet, humid weather. There are various types that affect brassicas, as well as onions. In brassicas the mould shows as a yellowish patch on the top of the leaf, with a corresponding whitish patch on the underside. If you notice these symptoms, destroy all infected leaves and improve air circulation around the plant. In onions, downy mildew affects the leaves first and can later spread to the bulbs. The leaf tips turn yellow and then blotches appear, which develop a furry white down. These patches eventually turn purplish-brown. Dig up and destroy infected crops.

Halo blight

This is a bacterial infection affecting French and runner beans. It is transmitted through the seed, and spreads from plant to plant via water droplets. Symptoms are often noticeable soon after germination, as affected seeds can produce brown, wrinkly seed leaves. As the plant grows, its leaves develop a dark spot in the centre of a paler ring or halo. The disease can usually be controlled by pulling off affected leaves when you see them or pulling up whole plants if badly stricken.

Honey fungus

This serious fungal disease can affect many species of trees and shrubs, including apple trees, and some fruit, including rhubarb and strawberries. Look out for white growths under the bark at the base of a tree or at the base of rhubarb stems, for example. There will also be tell-tale white threads in the soil, caused by the spreading fungus, and often confused with small tree roots. Dig up and burn infected plants; there is no cure.

Leek rust

Leek rust is a common infection, manifesting as orange or brownish spots on leaves and stems. The disease is often mild, in which case the crop will still be edible, but in some severe cases the leaves will die back, rendering the plant useless. To prevent this fungal infection give the plants plenty of space to help air circulation.

Mosaic virus

This is a common disease of cucumbers, courgettes, squashes, tomatoes and other vegetables, as well as some ornamental plants. It usually appears as yellowing, dying leaves and distorted fruit. It is spread by sap-sucking aphids, so effective control of these is the best preventative measure. Encourage hoverflies and ladybirds and use insecticidal soap.

Onion stem rot

Also known as neck rot, this is a disease that shows only after the bulbs are stored. The onions appear healthy in growth, but after 8–12 weeks in store they turn soft, with a brownish-black discoloration beneath the outer layers. It can also affect shallots. Check onions in store regularly and remove and destroy any that show signs of infection. Make sure that you allow onions to dry properly after lifting and before storing.

Onion white rot

This is the most widespread and serious disease of the allium family. The first sign in onions is wilting leaves. The bulbs start rotting from the base and are covered with a white, fluffy mould. It is a fungal infection, and spores can remain in the soil for 15 years, sometimes staying dormant until a host crop has been planted. Remove infected plants immediately and destroy, and avoid planting any members of the onion family in this section of the ground for as long as you possibly can.

Parsnip canker

This fairly common disease is unique to parsnips. It is a fungal infection, which usually enters through a scar or wound, and it turns the top or 'shoulders' of the parsnip brown. They then start to rot. There is no cure, but liming the soil to make it less acidic can help.

Powdery mildew

This fungal infection can occur on a range of fruit and vegetables, including strawberries, raspberries, cucumber, marrows and leaf vegetables, such as spinach, particularly when the roots are too dry. A dry, whitish powder spreads over leaves and shoot tips in summer, which can stunt and distort a plant's growth. Keeping plants well watered can help, and adding a thick mulch in spring can prevent the roots from drying out.

Reversion disease

This is a virus that is specific to blackcurrants and there is no cure. Spread by tiny mites, it affects the buds, which swell up (giving the disease its other name, big bud). Instead of going on to produce fruit in the normal way, the buds shrivel and dry up, and fruit production is severely limited. An early sign that the bush has reversion are the flowers, which appear bright pink and hairless (on healthy plants the flowers should be downy and greyish-pink). Unfortunately, there is no cure, so affected plants should be dug up and replaced with certified disease-free stock.

Spur blight

This fungal infection affects young raspberry canes in late summer. The first signs are purplish patches, usually around the leaf bud, which turn darker, then whitish. It will not kill the plant, but fruit production will be reduced the following year. It can be controlled by ensuring that the plants are not overcrowded, so thinning canes in autumn is important.

Violet root rot

Violet root rot can affect a range of vegetables, particularly root crops, such as potatoes, carrots, parsnips and turnips. It can also affect asparagus and strawberries. Tiny threads of purple fungi cover the roots and crowns of plants, causing yellowing leaves and stunted growth. It occurs most often in acid soils and, as with other fungal problems, in wet, humid weather. Affected crops should be destroyed.

resources

Seed suppliers

DT Brown
Tel: 0845 166 2275
Website: www.dtbrownseeds.co.uk
Well known supplier, good range of
vegetables and flowers.

Chiltern Seeds
Tel: 01229 581137
Website: www.edirectory.co.uk/
chilternseeds
Wide range of seeds, great choice and
some unusual varieties.

Dobies
Tel: 0870 112 3623
Website: www.dobies.co.uk
Good range of seed as well as
vegetable plugs.

Thomas Etty Esq.
Tel: 01963 359202
Website: www.thomasetty.co.uk
Old-fashioned supplier of heritage seed.
Lovely catalogue with line drawings and
historical snippets.

Mr Fothergills
Tel 0845 166 2511
Website: www.fothergills.co.uk
Very well established seed company
with a huge range.

Kings Seed
Tel: 01376 570000
Website: www.kingsseeds.com
Good range of excellent quality seed.

Kokopelli Organic Seeds
Website: www.organicseedsonline.com
A web-based company selling a
wonderful variety of heritage vegetable
seed.

Marshalls Seeds
Tel: 01480 443390
Website: www.marshalls-seeds.co.uk
Specifically for fruit and vegetables,
excellent on-line ordering.

Organic Gardening Catalogue
Tel: 0845 130 1304
Website: www.organiccatalog.com
Organic seed (both heritage and modern
varieties) plus organic fertilisers and
sundries.

W. Robinson & Son
Tel: 01524 791210
Website: www.mammothonion.co.uk
For those wanting to grow prize
vegetables.

Seeds of Italy
Tel: 0208 427 5020
Website: www.seedsofitaly.com
UK distributor for Franchi, Italy's
oldest seed company. Fabulous range
of old Italian varieties.

Suttons Seeds
Tel: 0870 220 2899
Website:
www.suttons-seeds.co.uk
Very well known company offering huge
range of vegetables, herbs and fruit, as
well as flowers.

Thompson & Morgan
Tel: 01473 695200
Website:
www.thompson-morgan.com
Highly regarded seed company with a
comprehensive list of vegetable seed.

Unwins Seeds
Tel: 01480 443395
Website: www.unwinsdirect.co.uk
Well known seed supplier.

Vida Verde
Tel: 01239 821107
Website: www.vidaverde.co.uk
A small-scale business offering small
range of very good varieties, all specially
selected for performance and flavour.

Fruit suppliers

Blackmoor Nurseries
Blackmoor Liss
Hampshire GU33 6BS
Tel: 01420 473576
Website: www.blackmoor.co.uk
Good range of fruit trees and soft fruit.
Visit the nursery or order online.

Keepers Nursery
Gallants Court
East Farleigh
Maidstone
Kent ME15 0LE
Tel: 01622 726465
Website: www.keepersnursery.co.uk
One of the leading fruit tree specialists
in the UK, with over 600 varieties,
from apples and pears to soft fruit.

It is worth a visit to the nursery or it is possible to order online.

❋ Ken Muir
Rectory Road
Weeley Heath
Clacton-on-Sea
Essex CO16 9BJ
Tel: 01255 830181
Website: www.kenmuir.co.uk
Ken Muir is a well known fruit specialist. The nursery is open to visit, or order online. The website includes articles on growing different fruit.

❋ Reads Nursery
Hales Hall
Loddon
Norfolk NR14 6QW
Tel: 01508 548395
Website: www.readsnursery.co.uk
Excellent quality fruit trees and soft fruit. Good website with useful training information. Visit the nursery or order online.

Associations

National Society for Allotment and Leisure Gardeners
O'Dell House
Hunters Road
Corby
Northants NN17 5JE
Tel: 01536 266576
Email: natsoc@nsalg.org.uk
Website: www.nsalg.org.uk
The national body representing allotments and allotmenteers throughout the country. Offers advice, information and a seed scheme. Membership £15 a year.

National Vegetable Society
5 Whitelow Road
Heaton Moor
Stockport SK4 4BY
Tel: 0161 442 7190
Website: www.nvsuk.org.uk
For those interested in growing vegetables for competitions and shows.

Royal Horticultural Society
80 Vincent Square
London SW1P 2PE
Tel: 0845 260 5000
Website: www.rhs.org.uk
Britain's leading gardening charity, with excellent resources online, an unrivalled horticultural library, and an excellent monthly journal. Membership starts at £42 per year.

Organic gardening

Garden Organic
Ryton Organic Gardens
Coventry
Warwickshire CV8 3LG
Tel: 024 7630 3517
Website: www.gardenorganic.co.uk
Formerly the HDRA, this is Britain's national charity for organic growing. It offers great resources including advice and information, online leaflets and a heritage seed library. Membership costs £26 per year for an individual and £30 for a family.

The Soil Association
Bristol House
40-56 Victoria Street
Bristol BS1 6BY
Tel: 0117 314 5000
Website: www.soilassociation.org
Britain's leading environmental charity promoting sustainable, organic farming.

Useful websites

❋ Allotments UK ❋
Website:
www.allotments-uk.com
Excellent resource with numerous useful links and a very lively forum making informative and interesting reading.

❋ Freecycle.org ❋
Website: freecycle.org
International initiative promoting the recycling of household goods – people have been known to get greenhouses for free this way! Visit the website to find local branch.

index

acknowledgments and credits

This book is dedicated to allotmenteers everywhere, and to my children, Charlie and Oliver. Clare Foster

First published in Great Britain in 2007 by Cassell Illustrated, a division of Octopus Publishing Group Ltd.
2-4 Heron Quays, London E14 4JP

Text copyright © 2007 Clare Foster
Design and layout © 2007 Octopus Publishing Group Ltd.
Reprinted 2008
The moral right of Clare Foster to be identified as the author of this Work has been asserted in accordance with the Copyright, Designs and Patents Act of 1988.

A CIP catalogue record for this book is available from the British Library.

ISBN-13: 978-1-844035-60-1

10 9 8 7 6 5 4 3 2 1

Designed by Abby Franklin
Illustrations by Matt Windsor
Publishing Manager Anna Cheifetz
Art Director Auberon Hedgecoe
Project Editor Joanne Wilson
Edited by Lydia Darbyshire
Index by Michèle Clarke
Printed in China

All photography: © Octopus Publishing Group Limited/ Francesca Yorke apart from the following:

Alamy/Arco images 158, 160 right; /blickwinkel 61, 139, 160 bottom centre; /Christopher Burrows 162 top left; /GardenWorld Images/Dave Bevan 105, 169 centre; /Daniel L. Geiger/SNAP 162 centre left; /Nic Hamilton 163 right; /Holt Studios International Limited/Nigel Cattlin 124, 159 left, 159 right, 160 top centre, 160 left; /Niall McDiarmid 104 left; /Duncan McNeill 83 bottom; /The Garden Picture Library/ Howard Rice 106; /Maciej Wojtkowiak 34 bottom right.
Corbis/Hulton-Deutsch Collection/Hans Bauman 4 top left, 6.
Frank Lane Picture Agency/Nigel Cattlin 164 left, 164 right, 164 centre, 165 left, 166 left, 167, 168, 169 left; /Andrew Linscott 166 centre.
Mark Veznaver/Twigs 97 top left, 101 bottom right, 109 centre, 113 top, 114, 115, 128, 134, 154 left, 155 top right, 156 top left, top right, 157 top left, bottom centre 168 centre.
Abby Franklin 142, 146 bottom.
Octopus Publishing Group Limited/Jerry Harpur 151; Howard Rice 102; /John Simms 163 centre, 165 right.